Mineral Beneficiation

T0228272

Mineral Beneficiation

A Concise Basic Course

D.V. Subba Rao

Head of the department of Mineral Beneficiation
S.D.S. Autonomous College
Andhra Pradesh, India

CRC Press
Taylor & Francis Group
Boca Raton London New York

CRC Press is an imprint of the
Taylor & Francis Group, an **informa** business

A BALKEMA BOOK

First published 2011 by CRC Press/Balkema

Published 2019 by CRC Press
Taylor & Francis Group
6000 Broken Sound Parkway NW, Suite 300
Boca Raton, FL 33487-2742

© 2011 Taylor & Francis Group, LLC
CRC Press is an imprint of the Taylor & Francis Group, an informa
business

First issued in paperback 2019

No claim to original U.S. Government works

ISBN-13: 978-0-367-45225-4 (pbk)
ISBN-13: 978-0-415-88228-6 (hbk)

This book contains information obtained from authentic and highly regarded sources. Reasonable efforts have been made to publish reliable data and information, but the author and publisher cannot assume responsibility for the validity of all materials or the consequences of their use. The authors and publishers have attempted to trace the copyright holders of all material reproduced in this publication and apologize to copyright holders if permission to publish in this form has not been obtained. If any copyright material has not been acknowledged please write and let us know so we may rectify in any future reprint.

Except as permitted under U.S. Copyright Law, no part of this book may be reprinted, reproduced, transmitted, or utilized in any form by any electronic, mechanical, or other means, now known or hereafter invented, including photocopying, microfilming, and recording, or in any information storage or retrieval system, without written permission from the publishers.

For permission to photocopy or use material electronically from this work, please access www.copyright.com (http://www.copyright.com/) or contact the Copyright Clearance Center, Inc. (CCC), 222 Rosewood Drive, Danvers, MA 01923, 978-750-8400. CCC is a not-for-profit organiza- tion that provides licenses and registration for a variety of users. For organizations that have been granted a photocopy license by the CCC, a separate system of payment has been arranged.

Trademark Notice: Product or corporate names may be trademarks or registered trademarks, and are used only for identification and explanation without intent to infringe.

Visit the Taylor & Francis Web site at
http://www.taylorandfrancis.com

and the CRC Press Web site at
http://www.crcpress.com

Library of Congress Cataloging-in-Publication Data

Subba Rao, D.V.

Mineral beneficiation : a concise basic course / D.V. Subba Rao.
p. cm.
Summary:"Mineral Beneficiation or ore dressing of run-of-mine ore is an upgrading process to achieve uniform quality, size and maximum tenor ore through the removal of less valuable material" -- Provided by publisher.
ISBN 978-0-415-88228-6 (hardback) -- ISBN 978-0-203-84789-3 (ebook)
1. Ore-dressing I. Title.

TN500.S87 2011
622'.7--dc22

2011009931

Typeset by Vikatan Publishing Solutions (P) Ltd, Chennai, India

Dedication

Late Seth Sriman Durgaprasadji Saraf
Founder, M/s Ferro Alloys Corp. Ltd, Shreeramnagar, A.P. India

who has given me the great opportunity of serving
students of Mineral Beneficiation in particular
and the Mineral Engineering Profession in general

Tributes to

Sri A.L. Mohan
Director, Infronics Systems India Ltd., Hyderabad, A.P. India

who lay the foundation for my professional growth

Contents

Preface

This book is the essence of thirty years of my experience as a teacher of undergraduate students of Mineral Beneficiation and ten years as a trainer of working Engineers of Kutch minerals, Essar Steels, Tega Industries, and Trimex Sands. During the first fifteen years, I experienced the difficulty of injecting the basic principles of Mineral Beneficiation into the minds of the students. I consulted with my colleagues and many engineers working in mineral beneficiation plants in search of a way out. I found that the explanation of basic principles given in various books available on the market lacks clarity in concepts because they were not written for this purpose.

Then a thought came into my mind to prepare notes exclusively on basic principles. An attempt was made in 1994, when I prepared the notes and conducted a short term course for my own students where all the basic principles were explained in just four days. The response was very good. During later years, the notes were revised by adding few additional points and conducting the course for five days for students as well as working engineers. The response was very encouraging.

Many engineers of mineral industries went through my notes, courtesy of my former students, and they were well appreciated. It was suggested that I should publish the book for the benefit of the Mineral Industries at large. This led me to write this book.

It is hoped that this small book will be of great use, not only for beginners but also for working engineers, erection engineers, designers, researchers and those who attend an interview at all levels. Any suggestions for improvement of this book will be appreciated, acknowledged and implemented in the right spirit.

D.V. Subba Rao
Head of the department of
Mineral Beneficiation
S.D.S.Autonomous College
Shreeramnagar,
Garividi – 535 101
Vizianagaram District
Andhra Pradesh
India
e-mail: dvsubbarao3@rediffmail.com

Foreword

I have great pleasure in writing the foreword for this book. Mr. D.V. Subba Rao and I are professional colleagues and have known each other for several years. His passion for Mineral Processing, in general, and teaching the fundamentals of the subject, in particular, are evident every time he steps into the classroom.

I have had the opportunity to see his teaching in action on several occasions. In fact, I attended a full training course given by him to the engineers of Tega Industries limited, Kolkata. The way the participants responded to him is amazing. He has an uncanny ability to break down complex concepts into easy-to-understand lessons.

In his book "COAL – ITS BENEFICIATION" (Subba Rao, 2003) he presents the concepts in simple way and includes lots of exercises to bring all the relevant concepts into one book to help students to grasp the principles and practicing engineers to apply the same for plant improvement studies.

The present book is his second book. I applaud the author for creating such a book, trying to bring out the fundamentals in the area of Mineral Beneficiation. In this book, Mr. Subba Rao has included everything he has learnt, whilst teaching and conducting short courses for plant engineers, and created an easy-to-understand way of presentation with illustrations which will provide proper fundamentals of the subject not only to the students but also to the practicing engineers. Throughout the book, an emphasis on making the mineral engineering a specialized subject in its own right is the central theme. I am sure that this book will win the approbation of students, practicing engineers, researchers, and all associated with mineral development including legends in the subject. I compliment Mr. Subba Rao for his dedicated service to the field of Mineral Processing.

If you are a student or practicing engineer searching for the fundamentals of Mineral Beneficiation/Processing to learn in the right way, this book is for you.

Dr. T.C. Rao
Formerly Director
Regional Research Laboratory, Bhopal

Professor and Head of the Department of
Fuel and Mineral Engineering
Indian School of Mines, Dhanbad

Acknowledgements

I am grateful to Dr. T.C. Rao, former Director, Regional Research Laboratory, Bhopal, and Professor and Head of the Department of Fuel and Mineral Engineering, Indian School of Mines, Dhanbad for continuous encouragement in writing this book.

I sincerely thank Prof. Kal Sastry, University of California, for his comments and suggestions on the contents of this book.

I am indebted to my colleagues Sri Y. Ramachandra Rao and Sri K. Satyanarayana, who helped me in various ways, including subject discussions and critical analysis and Sri A.K. Shrivastav and Sri K. Ganga Raju for their precious and supreme services rendered in bringing out this book.

I am thankful to FACOR management, in particular Sri R.K. Saraf (CMD), Dr. V. Subba Rao (Principal) and several other colleagues, both teaching and non-teaching, of S.D.S. Autonomous College for extending their cooperation in preparing this book. Special thanks go to Sri C. Raghu Kumar, one of my former students, who has taken lot of pains to help with this book.

It has been a pleasure to work with Taylor & Francis group and the cooperation of the editorial and production staff is much appreciated.

Finally, my deepest gratitude to my wife Mrs. Krishna Veni and daughters, Mrs. Radha Rani and Ms. Lalitha Rani for their unfailing emotional support in bringing out this book.

I gratefully acknowledge the following organizations for permitting me to use respective photographs:

Jayant Scientific Industries, Mumbai Maharashtra, India	Table Model Sieve Shaker Ro-tap Sieve shaker
Mogensen, Grantham, Lincolnshire, England	Divergator Vibrating grizzly
Deister Concentrator LLC, USA	Sieve Bend
Qingzhou Yuanhua Machinery Manufacture Co. Ltd., China	Trommel
Robert Cort Ltd., Reading, England	Vibrating Screen

FLSmidth Pvt Ltd, Kelambakkam, Tamilnadu, India	Cut section of Fuller-Traylor Gyratory Crusher
Pennsylvania Crusher Corporation, USA	Bradford Breaker
Metso Minerals Industries, Inc	Cut section of Cylindroconical Ball Mill
www.mine-engineer.com	Cut section of Cylindrical Ball Mill
Outotec (USA) Inc., Jacksonville,	Humphrey Spiral

List of tables

List of figures

Chapter 1

Introduction

The Earth's **Crust** is the topmost solid layer of the Earth which has a thickness of 30–35 km in the continents and 5–6 km in the oceans (the **Mantle** and the **Core** being the other two inner parts of the earth). According to *F.W. Clarke*, the abundance of Chemical elements in the earth's crust is shown in Table 1.1.

The chemical elements that occur in the earth's crust in compound forms are known as minerals. However, certain elements like Gold, Silver, Platinum and Copper often occur in native form.

The use of minerals has been instrumental in raising the standard of living of mankind. The sophisticated world of today is largely the result of the enlarged use of minerals for various purposes. All engineering and structural materials, machinery, plants, equipments and anything from a pin to a plane are manufactured from metals extracted from minerals. Some minerals form the starting point for basic industries like cement, fertilizer, ceramic, electrical, insulating, refractory, paint and abrasive materials and a host of chemicals. There is not a single industry which can do without minerals or their products. Minerals, thus, form part and parcel of our daily life.

Table 1.1 Abundance of chemical elements in the Earth's crust.

Chemical elements	Percentage
Oxygen	46.46
Silicon	27.61
Aluminium	8.07
Iron	5.06
Calcium	3.64
Sodium	2.75
Potassium	2.58
Magnesium	2.07
Titanium	0.62
Hydrogen	0.14
Phosphorus	0.12
Manganese	0.09
Carbon	0.09
Sulphur	0.06
Barium	0.04
Flourine	0.03
Strontium	0.02
All other elements	0.50

1.1 MINERALS

As defined by Dana, a well known physicist:

Mineral is a substance having definite chemical composition and internal atomic structure and formed by the inorganic processes of nature [1]

Minerals are broadly classified into two types:

1 Metallic minerals.
2 Non-metallic minerals.

Metallic minerals are the minerals from which a metal is extracted. A few metallic minerals, their chemical formulae, metal extracted and the percent metal present in the mineral are shown in Table 1.2.

The minerals of Uranium and Thorium are also called **atomic minerals.**

Non-metallic minerals are the minerals used for industrial purposes for making cement, refractories, glass and ceramics, insulators, fertilizers etc. These minerals are also called **industrial minerals.** Metals are not extracted from these minerals. Some metallic minerals are also used for industrial purposes like Bauxite, Chromite and Zircon for the refractory industry, Pyrolusite for dry battery cells and Ilmenite for the pigment industry, etc. A few non-metallic minerals with their chemical formulae are shown in Table 1.3.

The third type, **coal,** is considered a mineral and is sometimes spoken of as mineral coal in trade, industry and legal affairs. But in a restricted technical sense, coal is not a mineral. It is organic in composition and formed from decaying vegetation and mineral matter. As it is a useful part of the earth's crust and requires treatment before use, it can be classed as third type of special significance.

Table 1.2 Metallic minerals.

Mineral	Chemical formula	Metal extracted	% metal
Hematite	Fe_2O_3	Iron	69.94
Magnetite	Fe_3O_4	Iron	72.36
Bauxite	$Al_2O_3 \cdot 2H_2O$	Aluminium	39.11
Braunite	$3Mn_2O_3 \cdot MnSiO_3$	Manganese	63.60
Pyrolusite	MnO_2	Manganese	63.19
Chromite	$FeO \cdot Cr_2O_3$	Chromium	46.46
Galena	PbS	Lead	86.60
Sphalerite	ZnS	Zinc	67.10
Chalcopyrite	$CuFeS_2$	Copper	34.63
Ilmenite	$FeO \cdot TiO_2$	Titanium	31.57
Rutile	TiO_2	Titanium	59.95
Zircon	$ZrSiO_4$	Zirconium	49.76
Pitchblende	U_3O_8	Uranium	84.80
Monazite	$(Ce,La,Th)PO_4$	Thorium	–

Table 1.3 Non-metallic minerals.

Mineral	Chemical formula
Andalusite	Al_2SiO_2
Apatite	$Ca_4(CaF)(PO_4)_3$
Asbestos (Crysotile)	$Mg_3Si_2O_5(OH)_4$
Baryte	$BaSO_4$
Bentonite	$(Ca\ Mg)O\ SiO_2\ (Al\ Fe)_2O_3$
Calcite	$CaCO_3$
Corundum	Al_2O_3
Diamond	C
Diaspore	$Al_2O_3H_2O$
Dolomite	$CaMg(CO_3)_2$
Feldspar	$(Na,K,Ca)\ AlSi_3O_8$
Fluorite	CaF_2
Garnet (Almandine)	$3FeO\ Al_2O_3\ 3SiO_2$
Graphite	C
Gypsum	$CaSO_4.2H_2O$
Kaolinite (China clay)	$H_4Al_2Si_2O_9$
Kyanite	Al_2SiO_5
Limestone	$CaCO_3$
Marble	Chiefly $CaCO_3$
Magnesite	$MgCO_3$
Mica (Muscovite)	$KAl_2(AlSi_3O_{10})(OH,F)_2$
Phosphate rock	$Ca_3(PO_4)_2$
Pyrite	FeS_2
Pyrophyllite	$H\ Al(SiO_3)_2$
Quartz	SiO_2
Sillimanite	$Al_2O_3\ SiO_2$
Talc	$H_2\ Mg_3(SiO_3)_4$
Vermiculite	$3\ MgO(FeAl)_2O_3\ 3SiO_2$

1.2   IMPORTANT TERMINOLOGY

Minerals do not occur singly in the earth's crust. They occur in association with several other minerals. The following important terminology is used in describing the mineral deposits and related terms.

Rock is an aggregation of several minerals as occurred in the earth's crust.
Ore is also an aggregation of several minerals from which one or more minerals can be exploited/separated at profit.

> All Ores are Rocks, but all Rocks are not Ores
> An Ore at one place may be a Rock at other place

Ore Minerals or **Valuable Minerals** are those minerals which contain an economically exploitable quantity of some metal or non-metal.
Gangue Minerals are usually the non-metallic minerals associated with ore minerals which are worthless as a source for that metal or otherwise. These are usually

unwanted, waste or useless minerals. These gangue minerals occasionally find use as source of by-products. For example, pyrite present in Lead and Zinc ores is a gangue mineral but it is separated as by-product for extraction of sulphur after the lead and zinc minerals are separated.

Ore Deposits are the natural deposits of ore minerals.

Ore is an aggregation of valuable and gangue minerals.

Simple Ore is one from which a single metal can be extracted. For example, only Iron is extracted from Hematite ore, Aluminium is extracted from Bauxite ore, Chromium is extracted from Chromite ore, etc.

Complex Ore is one from which two or more metals can be extracted. Lead, Zinc and Copper metals are extracted from Lead-Zinc-Copper Ore.

Metal Content of a mineral is generally expressed in percent of metal present in the mineral. It is calculated by taking the atomic weights of the elements present in the mineral.

Let us consider **Hematite (Fe_2O_3)**

Atomic weight of Iron = 55.85
Atomic weight of Oxygen = 16.00
Molecular weight of Hematite = $55.85 \times 2 + 16 \times 3 = 159.7$

$$\text{Percent Iron} = \frac{55.85 \times 2}{159.7} \times 100 = 69.94$$

Assay Value or **Tenor** is the percent metal, percent valuable mineral, or ounces precious metal per ton depending upon the type of ore involved.

Grade is a relative term used to represent the value of an ore.

High Grade Ore is an ore having a high assay value and **Low Grade Ore** is an ore having a low assay value. The Ore having an assay value between that of high and low value is called **Medium Grade Ore.**

Rich Ore and **Lean Ore** are the other terms of common usage where an ore with a high assay value is rich ore and an ore with low assay value is lean ore.

1.3   BENEFICIATION

Separation of the wanted part from the aggregation of wanted and unwanted parts by physical methods is termed as Beneficiation. Separation of rice from the mixture of rice and stones is the example known to everyone.

1.4   MINERAL BENEFICIATION

As defined by A.M. Gaudin

Mineral Beneficiation can be defined as processing of raw minerals to yield marketable products and waste by means of physical or mechanical methods in such a way that the physical and chemical identity of the minerals are not destroyed [2].

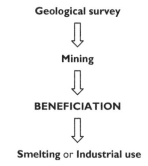

Figure 1.1 The major steps in processing of ores.

It follows that mineral beneficiation is a process designed to meet the needs of the consumer of minerals.

Run-of-mine Ore is an ore directly taken from the mine, as it is mined.

Figure 1.1 shows the successive major steps involved in processing the ores.

Geologists conduct a geological survey and estimate the ore reserves, their quality and tenor. Mining engineers mine the ore and bring it to the surface of the earth. Mineral Engineers beneficiate the ore to higher tenor. Thus beneficiated ore, if it is metallic ore, is smelted and the metal is extracted which is further utilized for the production of alloys. If the ore is non metallic, beneficiated ore is directly utilized for the production of various products like cement, fertilizers etc.

Smelting operation, for the extraction of a metal, requires:

- Uniform quality of the ore.
- Appropriate size of the ore.
- Minimum tenor of the ore.

Beneficiation of run-of-mine ore is done to achieve the above. The primary object of Mineral Beneficiation is to eliminate either unwanted chemical species or particles of unsuitable size or structure.

During beneficiation, much of the gangue minerals, usually present in large quantities in many ores, are eliminated or removed.

The benefits are:

1 Freight and handling costs reduced.
2 Cost of extraction (smelting) reduced.
3 Loss of metal in slag reduced.

By doing beneficiation, lean ores can be made technically suitable for extraction of metal. Mineral Beneficiation is usually carried out at the mine site. The essential reason is to reduce the bulk of the ore which must be transported, thus saving the transport cost.

The reasons for the increasing importance of Mineral Beneficiation are:

1   Reserves of good quality ore (high grade ore) are depleting day by day as much of such ore is continuously mined and utilized for extraction of metal and hence it is unavoidable to use low grade ores (which need beneficiation) for metal extraction to meet the demands.
2   In order to use un-mined reserves of a particular mine, switching over from **Selective mining** to a cheaper mining method, **Bulk mining,** is found to be more economical wherein beneficiation is a must.
3   When compared to metallurgical processes Mineral Beneficiation is inexpensive.

1.5   MINERAL BENEFICIATION OPERATIONS

The following are some of the synonymous terms used for Mineral Beneficiation:

Mineral Dressing
Mineral Processing
Ore Dressing
Ore Processing
Ore Preparation
Ore Concentration
Ore Upgradation
Ore Enrichment
Milling

Even though the individual terms have their own significance and meaning, it can be taken that all the terms are alike as far as simple understanding is considered.

The principal steps involved in beneficiation of Minerals are:

1   **Liberation:** Detachment or freeing of dissimilar particles from each other i.e. valuable mineral particles and gangue mineral particles.

Operations: Crushing
                        Grinding

2   **Separation:** Actual separation of liberated dissimilar particles i.e., valuable mineral particles and gangue mineral particles.

Operations: Gravity concentration
                          Heavy Medium Separation
                          Jigging
                          Spiraling
                          Tabling
                        Flotation
                        Magnetic separation
                        Electrical separation
                        Miscellaneous operations like hand sorting

Supporting Operations: Preliminary washing
                                           Screening
                                           Classification
                                         Thickening
                                         Filtration
                                         Handling of materials
                                             Storage
                                           Conveying
                                           Feeding
                                           Pumping
                                       Pneumatic and Slurry transport

Supporting operations (one or the other) are essential operations of any plant without which no plant exists. A road metal crusher, for example, can perform its job only when proper arrangements are made to feed the metal to the crusher and to convey the crushed metal for separating it into different sizes.

1.6 UNIT OPERATIONS

The operations conducted on any material that involve physical changes are termed as **Unit Operations**. The various operations performed for the beneficiation of minerals are all unit operations as the changes in these operations are primarily physical. The basic principles involved in Unit Operations are independent of the material treated. In designing a treatment method, it is essential to recognize the unit operations to be performed.

1.7 TECHNOLOGY

The term technology is used in the sense of the application of technical skills along with the economic justification of the operations. A sound knowledge of basic sciences such as chemistry, physics and mathematics as well as engineering crafts required in the handling of large tonnages of ore and fluids, is of prime importance in the study of this technology. Experience and judgment play an important part in the application of theoretical principles.

Chapter 2

Ore formation, identification and analysis

2.1 ORE FORMATION

Ores in the earth's crust have been formed by several processes. These ore-forming processes have been classified from time to time by several workers. However, in 1950, Bateman proposed a classification of the processes as shown below [3]:

1 Magmatic concentration.
2 Sublimation.
3 Contact metasomatism.
4 Hydrothermal processes.
5 Sedimentation.
6 Evaporation.
7 Residual and mechanical concentration.
8 Oxidation and supergene enrichment.
9 Metamorphism.

According to the manner of formation, ore deposits are divided into three great types as given below:

1 Igneous.
2 Sedimentary.
3 Metamorphic.

2.2 IDENTIFICATION OF MINERALS

The following are the some of the physical properties of minerals through which minerals can be identified before they are put to use:

1 Characters dependent upon light

    a Colour
    b Streak
    c Lustre
    d Transparency

  e Phosphorescence
  f Fluorescence

2 Taste, odour and feel
3 State of aggregation

  a Form
  b Habit
  c Pseudomorphism, Polymorphism and Polytipism
  d Cleavage
  e Fracture
  g Hardness
  h Tenacity

4 Specific gravity (density)
5 Magnetic susceptibility
6 Electrical conductivity
7 Radioactivity
8 Surface property

The identification of minerals by their physical properties is termed as Megascopic Identification. Minerals are also identified by their optical properties under a microscope. Transparent minerals are identified under a Petrological or Mineralogical microscope whereas opaque minerals are identified under an Ore microscope. This microscopic examination, carried out for thorough understanding of the mineralogy of ore, determine the mineral species present in the ore and their relative abundance.

  Texture of mineral occurrences is an important property useful for separation of valuable minerals from their ores. Textures are mainly of three types:

Fine-grained <1 mm
Medium-grained 1–5 mm
Coarse-grained >5 mm

2.3   MINERAL ANALYSIS

Minerals are analyzed by conventional chemical analysis. Two types of chemical analysis are:

1 **Qualitative Analysis,** in which elements present in the sample are identified.
2 **Quantitative Analysis,** in which the quantity of elements, or compounds, present in the sample is estimated.

  In a few instances it may be possible to calculate the mineral proportions of a sample specimen from the results of a chemical analysis. More usually, chemical analysis can, at best, only provide a rough estimate of the mineral content of an ore or plant product. The combined information on mineral identities, mineral compositions and mineral proportions, will establish how the various chemical elements are partitioned (or shared) among the various minerals. This information can then be used to estimate both the quantities and the qualities of the various products that may be obtained from an ore.

Several varieties of instrumental methods are also available for analysis of ores and minerals.

The three following points are to be noted carefully in case of metallic ores:

1   The grade or quality of an ore is represented by its **metal** content.
2   The **minerals** are separated during beneficiation.
3   The **metal** is extracted through metallurgical operations.

Chapter 3

Sampling

A Mineral Beneficiation plant costs thousands of dollars to build and operate. The success of the plant relies on the assays of a few small samples. Representing large ore bodies truly and accurately by a small sample that can be handled in a laboratory is a difficult task. The difficulties arise chiefly in collecting such small samples from the bulk of the material.

The method or operation of taking the small amount of material from the bulk is called **Sampling**. It is the art of cutting a small portion of material from a large lot. The small amount of material is called **Sample** and it should be representative of the bulk in all respects (in its physical and chemical properties). More precisely, sampling can be defined as the operation of removing a part, convenient in quantity for analysis, from a whole which is much greater, in such a way that the proportion and distribution of the quality to be tested are the same in both the sample and the whole.

Sampling is a statistical technique based on the theory of probability. The first and most obvious reason for sampling is to acquire information about the ore entering the plant for treatment. The second is to inspect its condition at selected points during its progress through the plant so that comparison can be made between the optimum requirements for efficient treatment and those actually existing, should these not coincide. The third is to disclose recovery and reduce losses.

The prerequisite for the development of a satisfactory flowsheet is the acquisition of a fully representative ore sample, even though, in respect of a new ore-body, this sample may have to be something of a compromise. A bad sample will result in wastage of all test work and can lead to a completely wrongly designed mill.

A sample can be taken from any type of material dry, wet or pulp. But, in each case, the method of sampling and the apparatus necessary for them are different.

A sample is collected from huge lot of dry material in stages. At first, a large quantity sample is collected from a lot, known as **primary sample** or **gross sample**, by means of various types of sampling equipment such as mechanical or hand-tool samplers using appropriate sampling methods and techniques.

The two methods used to obtain a gross sample are **Random sampling** and **Systematic sampling**. The various hand-tool samplers used are Drill, Shovel, Scoop, Auger, Pipe and Slot samplers. The gross sample is reduced to a quantity that can be handled with ease by **alternate shoveling** or **fractional shoveling** in stages depending upon the quantity of the gross sample. It is essential that the gross sample be thoroughly mixed before reduction in order to obtain a representative sub-sample or laboratory sample.

Such reduced samples are called **secondary sample** and **ternary sample** depending upon the number of stages used. Figure 3.1 shows the stages of sampling. Reduction of this reduced sample to a quantity necessary for analysis, known as **final sample** or **test sample**, is called **sample preparation**. It is the process of reducing the quantity by splitting.

Sample preparation is done by **Coning and quartering** or by using **paper cone splitter, riffle splitter, rotary cone splitter, rotary table splitter** or **micro splitter.** The sampler's knowledge, experience, judgment and ability are of greater value because instructions cannot cover every point or combination of circumstances encountered on each preparation.

When it is required to collect samples from streams of solids and/or pulps, **manual** or **mechanical sample cutters** are employed to cut and withdraw small quantities from a stream of traveling material at predetermined frequencies and speeds to form a gross sample. The sample cutter should travel across the material stream and intersect the stream perpendicular to the flow so that the material from the entire width of the stream is collected. The cutter width should be 3 times the top size of the particle and should travel at constant speed. A common falling ore sampling device is the Vezin sampler. Sampling devices called **poppet valves** are used for pulp sampling in pipes. These are typically used in pipes where the flow is upward.

The procedure to be adopted for taking a sample and the amount of the sample depends on the size of the original material, particle size of the material, the method of sampling and the purpose for which the sample is taken. Table 3.1 is one of the early sampling studies that proposed to relate the particle size of the material being sampled to the sample size required for a representative sample.

The basis for the sample for this theory was 100 tons of ore. As one can see, the finer the material being sampled, the smaller the size of sample required. Owing to the statistical fact that the finer particles have many more individual particles per pound than the coarser particles do and, since ore is made up of many different materials,

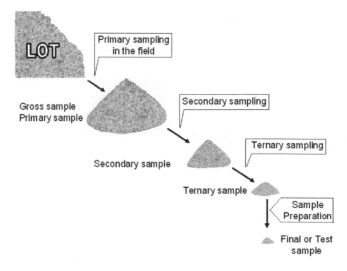

Figure 3.1 Sampling process.

Table 3.1 Particle size and minimum weight of the sample.

Particle size inches	Minimum weight of the sample, pounds
0.04	0.0625
0.08	0.5
0.16	4
0.32	32
0.64	256
1.25	2048
2.50	16348

the finer particles are much more likely to contain all of the individual elements of the whole sample.

When a sample is to be taken for chemical analysis to determine the assay value of the ore, the sample should be re-crushed sufficiently between each cutting down of the sample so that the ratio of the diameter of the largest particles to the weight of the sample to be taken shall not exceed a certain safe proportion. It is to be noted that no amount of mixing and careful division can make the sample and reject alike in value when the lot before division contains an uneven number of large high grade ore particles.

In a process plant, or mine, the preferred method of obtaining a sample, is from a moving stream, such as a conveyor belt, a slurry pipe line, or perhaps from a chute that a stream of ore flows through. The material collected each time is called an **increment**. How the sample is obtained, the number of increments and the size of each increment, will often determine the degree of probability that a sample is indeed representative. The **sampling ratio**, which is defined as the ratio of the weight of the sample taken by the sampling system to the weight of the lot from which that sample is taken, is the most important indicator of the performance of the sampling system.

When the sample is taken, and subjected to analysis, there exists some chance of error in a single sample. One must take the number of samples to reduce the error and to keep the overall error within the tolerable working limits.

Chapter 4

Size

Size of the particle is an important consideration in Mineral Beneficiation because of the following main reasons:

- Energy consumed for reducing the size of the particles depends on size.
- Size of the particles determines the type of size reduction equipment, beneficiation equipment and other equipment to be employed.

The size of the particle of standard configuration like sphere and cube can easily be specified. For example, the size of a spherical particle is its diameter (d) and that of a cubical particle is the length of its side (l) as shown in Figure 4.1.

As the mineral particles are irregular in shape, it is difficult to define and determine their size. A number of authors have proposed several empirical definitions to particle size. Feret [4], in 1929, defined the size of an irregular particle as the distance between the two most extreme points on the surface of a particle. Martin [4], in 1931, defined the length of the line bisecting the maximum cross-sectional area of the particle. During later years, the size of a particle is defined by comparison with a standard configuration, normally a spherical particle.

Equivalent size or **equivalent diameter** of an irregular particle is defined as the diameter of a spherical particle which behaves similar to an irregular particle under specified conditions.

Surface diameter is defined as a diameter of a spherical particle having the same surface area as the irregular particle.

Volume diameter is defined as the diameter of a spherical particle having the same volume as the irregular particle.

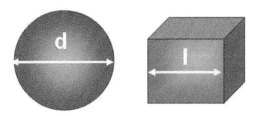

Figure 4.1 Spherical and cubical particles.

Figure 4.2 Test sieve.

It is obvious that each definition has its own limitations.

In mineral industry, the side of a square aperture through which a particle just passes is taken as the size of the particle even though little or no importance is given to its shape. Standard Test Sieves are used in the mineral industry to measure the size of the small and the fine particles, usually down to 74 microns.

Test Sieve is a circular shell of brass having an 8 inch diameter and being about 2 inch high as shown in Figure 4.2.

Sieve cloth is made of wire, woven to produce nominally uniform cloth apertures (openings). The sieve cloth is placed in the bottom of the shell so that material can be held on the sieve.

Aperture (or **Opening**) is a distance between two parallel wires.

Mesh number is the number of apertures per linear inch. Sieves are designated by mesh number.

Mesh size is the size of an aperture i.e. the distance between two parallel wires. As mesh number increases, mesh size decreases.

Sieve Scale is the list of successive sieve sizes used in any laboratory, taken in order from coarsest to finest.

Standard Sieve Scale is the sieve scale adopted for size analyses and general testing work to facilitate the interchangeability of results and data. In this standard sieve scale, the sizes of successive sieves in series form a geometric progression.

For a standard sieve scale, the reference point is 74 microns, which is the aperture of a 200 mesh woven wire sieve. The ratio of the successive sizes of the sieves in the standard sieve scale is $\sqrt{2}$, which means that the area of the opening of any sieve in the series is twice that of the sieve just below it and one half of the area of the sieve above it in the series.

In general, mesh number × mesh size in microns ≈ 15,000.

For closer sizing work the sieve ratio of $\sqrt[4]{2}$ is common.

The different standards in use are:

American Tyler Series
American Standards for Testing and Materials, ASTM E-11-01

Table 4.1 Comparison of test sieves of different standards.

TYLER			U.S.A. ASTM E-11-01		BRITISH B.S 410-2000		INDIAN I.S. 460-1962		FRENCH AFNOR NFX-11-501		GERMAN DIN 3310-1:2000	
Sieve designation mesh no.	Width of aperture mm	Mesh double tyler series	Sieve designation mesh no.	Width of aperture mm	Sieve designation mesh no.	Width of aperture mm	Sieve designation mesh no.	Width of aperture mm	Sieve designation mesh no.	Width of aperture mm	Sieve designation mesh no.	Width of aperture mm
-	-	-	-	-	-	-	-	-	38	5.00	-	5.00
4	4.75	-	4	4.75	3½	4.75	480	4.75	-	-	-	4.50
-	4.00	5	5	4.00	4	4.00	400	4.00	37	4.00	2E	4.00
6	3.35	-	6	3.35	5	3.35	340	3.35	-	-	-	-
-	-	-	-	-	-	3.15	320	3.18	36	3.15	-	3.15
-	2.80	7	7	2.80	6	2.80	280	2.80	-	-	-	2.80
8	2.36	-	8	2.36	7	2.36	240	2.39	35	2.50	-	2.50
-	2.00	9	10	2.00	8	2.00	200	2.00	34	2.00	3E	2.00
10	1.70	-	12	1.70	10	1.70	170	1.70	33	1.60	-	1.60
-	1.40	12	14	1.40	12	1.40	140	1.40	-	1.40	-	1.40
-	-	-	-	-	-	1.25	-	-	32	1.25	5	1.25
14	1.18	-	16	1.18	14	1.18	120	1.20	-	-	6	1.20
-	1.00	16	18	1.00	16	1.00	100	1.00	31	1.00	-	1.00
20	0.85	-	20	0.850	18	0.850	85	0.850	-	-	-	-
-	-	-	-	-	-	0.800	80	0.79	30	0.800	-	0.800
-	0.710	24	25	0.710	22	0.710	70	0.710	-	0.710	-	0.710
-	-	-	-	-	-	0.630	-	-	29	0.630	-	0.630
28	0.600	-	30	0.600	25	0.600	60	0.600	-	-	10	0.600
-	0.500	32	35	0.500	30	0.500	50	0.500	28	0.500	12	0.500
35	0.425	-	40	0.425	36	0.425	40	0.425	-	-	-	-
-	-	-	-	-	-	0.400	-	-	27	0.400	-	0.400
-	0.355	42	45	0.355	44	0.355	35	0.355	-	0.355	16	0.355
-	-	-	-	-	-	0.315	-	-	26	0.315	-	0.315
48	0.300	-	50	0.300	52	0.300	30	0.300	-	-	20	0.300
-	0.250	60	60	0.250	60	0.250	25	0.250	25	0.250	24	0.250
65	0.212	-	70	0.212	72	0.212	20	0.212	-	-	-	-

(Continued)

Table 4.1 (Continued).

TYLER Sieve designation mesh no.	TYLER Width of aperture mm	TYLER Mesh double tyler series	U.S.A. ASTM E-11-01 Sieve designation mesh no.	U.S.A. Width of aperture mm	BRITISH B.S 410-2000 Sieve designation mesh no.	BRITISH Width of aperture mm	INDIAN I.S. 460-1962 Sieve designation mesh no.	INDIAN Width of aperture mm	FRENCH AFNOR NFX-11-501 Sieve designation mesh no.	FRENCH Width of aperture mm	GERMAN DIN 3310-1:2000 Sieve designation mesh no.	GERMAN Width of aperture mm
–	–	–	–	–	–	0.200	–	–	24	0.200	30	0.200
–	0.180	80	80	0.180	85	0.180	18	0.180	–	0.180	–	0.180
–	–	–	–	–	–	0.160	–	–	23	0.160	–	0.160
100	0.150	–	100	0.150	100	0.150	15	0.150	–	–	40	0.150
–	0.125	115	120	0.125	120	0.125	12	0.125	22	0.125	50	0.125
150	0.106	–	140	0.106	150	0.106	10	0.106	–	–	–	–
–	–	–	–	–	–	0.100	–	–	21	0.100	60	0.100
–	0.90	170	170	0.090	170	0.090	9	0.090	–	0.090	70	0.090
–	–	–	–	–	–	0.080	–	–	20	0.080	–	0.080
200	0.075	–	200	0.075	200	0.075	8	0.075	–	–	80	0.075
–	–	–	–	–	–	0.071	–	–	–	0.071	–	0.071
–	0.063	250	230	0.063	240	0.063	6	0.063	19	0.063	–	0.063
–	–	–	–	–	–	0.056	–	–	–	0.056	110	0.056
270	0.053	–	270	0.053	300	0.053	5	0.053	–	–	–	–
–	–	–	–	–	–	0.050	–	–	18	0.050	–	0.050
–	0.045	325	325	0.045	350	0.045	4	0.045	–	0.045	120	0.045
–	–	–	–	–	–	0.040	–	–	17	0.040	–	0.040
400	0.038	–	400	0.038	400	0.038	3	0.038	–	–	130	–

British Standard Sieves, BSS 410-2000
French Series, AFNOR (Association Francaise de Normalisation)NFX 11-501
German Standard, DIN (Deutsches Institut fur Normung) 3310-1:2000

The Indian Standard (IS) sieves, however, follow a different type of designation. For an IS sieve, the mesh number is equal to its aperture size expressed to the nearest deca-micron (0.01 mm). Thus an IS sieve of mesh number 50 will have an aperture width of approximately 500 microns. Such a method of designation has the simplicity that the aperture width is readily indicated from the mesh number.

For most size analyses it is usually impracticable and unnecessary to use all the sieves in a particular series. For most purposes, alternative sieves are quite adequate. For accurate work over certain size ranges of particular interest, consecutive sieves may be used. Intermediate sieves should never be chosen at random, as the data obtained will be difficult to interpret. In general, the sieve range should be chosen so that no more than about 5% of the sample material it retained on the coarsest sieve, or passes the finest sieve. These limits may be lowered for more accurate work.

Table 4.1 shows the comparison of test sieves of different standards.

4.1 SIEVE ANALYSIS

It is a method of size analysis. It is performed to determine the percentage weight of closely sized fraction by allowing the sample of material to pass through a series of test sieves.

Closely sized material is the material in which the difference between maximum and minimum sizes is less.

Sieving can be done by hand or by machine. The hand sieving method is considered more effective as it allows the particles to present in all possible orientations on to the

Figure 4.3 Table model sieve shaker.

(Courtesy Jayant Scientific Industries, Mumbai).

Figure 4.4 Ro-tap sieve shaker.

(Courtesy Jayant Scientific Industries, Mumbai).

sieve surface. However, machine sieving is preferred for routine analysis as hand sieving is long and tedious. Table model sieve shaker and Ro-tap sieve shaker (Figures 4.3 and 4.4) are the two principal machines used in a laboratory for sieve analysis.

Owing to irregular shapes, particles cannot pass through the sieve unless they are presented in a favourable orientation, particularly with the fine particles. Hence there is no end point for sieving. For all practical purposes, the end point is considered to have been reached when there is little amount of material passing through after a certain length of sieving.

Sieving is generally done dry. Wet sieving is used when the material is in the form of slurry. When little moisture is present, a combination of wet and dry sieving is performed by initially adding water.

4.2 TESTING METHOD

The sieves chosen for the test are arranged in a **stack,** or **nest,** starting from the coarsest sieve at the top and the finest at the bottom. A pan or receiver is placed below the bottom sieve to receive the final undersize, and a lid is placed on top of the coarsest sieve to prevent escape of the sample.

The material to be tested is placed on the uppermost coarsest sieve and closed with lid. The nest is then placed in a Sieve Shaker and sieved for certain time. Figure 4.5 shows the sieve analysis at the end of the sieving.

The material collected on each sieve is removed and weighed. The complete set of values is known as **Particle Size Distribution** data. Particle size distribution refers

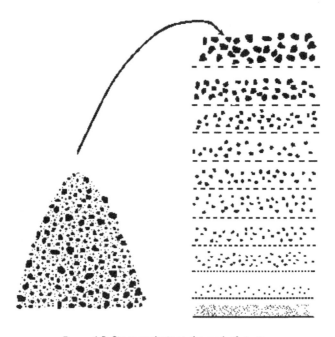

Figure 4.5 Sieve analysis at the end of sieving.

to the manner in which particles are quantitatively distributed among various sizes; in other words a statistical relation between quantity and size. Particle size distribution data is presented in a tabular form as shown in Table 4.2.

The weight percentages of the material retained on each sieve are to be calculated to form differential analysis. Cumulative weight percentage retained is obtained from differential analysis by adding, cumulatively, the individual differential weight percentages from the top of the table. Cumulative weight percentage passing is obtained by adding, cumulatively, the individual weight percentages from the bottom of the table.

All the fractions are fairly closely sized (except first fraction). Hence the size of the particles in each fraction may be calculated as arithmetic mean of the limiting sizes.

For example, the size of −14 + 22 mesh fraction is $\dfrac{1200 + 710}{2} = 955$ microns.

It means, the particles which pass through 14 mesh and retain on 22 mesh are having the mean size of 955 microns. Similarly the mean sizes of each fraction are to be calculated. Table 4.3 shows all values.

The average size of the material is determined by using the following simple arithmetic formula

$$\therefore \text{Average size} = \frac{100}{\sum \frac{w_i}{d_i}} \qquad\qquad 4.2.1$$

where w is the weight percent of the material retained by the sieve and d is the mean size of the material retained by the same sieve.

Table 4.2 Particle size distribution data from size analysis test.

Mesh number	Retained mesh size in microns	Weight of material gm
+14	1200	02.5
−14 + 22	710	18.0
−22 + 30	500	18.5
−30 + 44	355	21.0
−44 + 60	250	27.5
−60 + 72	210	36.0
−72 + 100	150	31.5
−100 + 150	105	26.0
−150 + 200	74	18.5
−200		50.5
		250.0

+ sign designates particles retained on that sieve.
− sign designates particles passed through that sieve.

Table 4.3 Calculated values for particle size distribution.

Mesh number	Retained mesh size microns	Mean size d_i microns	Weight gm	wt % retained w_i	Cum wt % retained	Cum wt % passing W
						100.0
+14	1200	1200	02.5	1.0	1.0	99.0
−14 + 22	710	955	18.0	7.2	8.2	91.8
−22 + 30	500	605	18.5	7.4	15.6	84.4
−30 + 44	355	427.5	21.0	8.4	24.0	76.0
−44 + 60	250	302.5	27.5	11.0	35.0	65.0
−60 + 72	210	230	36.0	14.4	49.4	50.6
−72 + 100	150	180	31.5	12.6	62.0	38.0
−100 + 150	105	127.5	26.0	10.4	72.4	27.6
−150 + 200	74	89.5	18.5	7.4	79.8	20.2
−200		37	50.5	20.2	100.0	
			250.0	100.0		

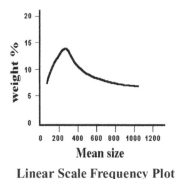

Mean size

Linear Scale Frequency Plot

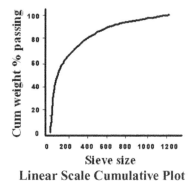

Sieve size

Linear Scale Cumulative Plot

Figure 4.6 Graphical presentation of particle size distribution data.

4.3 PRESENTATION OF PARTICLE SIZE DISTRIBUTION DATA

Particle size distribution data is best presented for use in the form of graphs (Figure 4.6). The simplest method is to plot a histogram of the weight percent of the material in the size interval against the size interval. When the size intervals are small enough, the histogram can be presented as a continuous curve taking the middle points of the histogram. In other words, a graph is plotted between the weight percent of the material as ordinate and the arithmetic mean size as abscissa. It is called a **linear scale frequency plot.** It gives the quantitative picture of the relative distribution of the material over the entire size range. In many cases, the data is more commonly plotted as cumulative weight percent passing versus actual size of opening. It is called a **linear scale cumulative plot.**

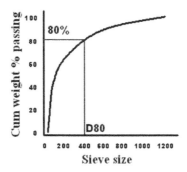

Figure 4.7 Plot for determination of 80% passing size D80.

The same graphs can also be drawn on semi-logarithmic graph paper for satisfactory spreading of the data on fine size region.

80% passing size (D80) is the size of the sieve at which 80% of the particles pass through that sieve. 80% passing size can be determined from the plot of cumulative weight percent passing versus sieve size as shown in Figure 4.7.

F80 is the 80% passing size of the feed material.
P80 is the 80% passing size of the product material.

80% passing size is used in all calculations to determine energy requirements for reducing the size of the particles by comminution equipment.

4.4 APPLICATIONS OF PARTICLE SIZE DISTRIBUTION DATA

1 Comparative efficiencies of comminution units by relating the work done and the product sizes can be studied.
2 Particle surface areas can be calculated from size analysis.
3 Power required to crush and/or grind an ore from a given feed size to a given product size can be estimated from size analyses of the feed and the products.
4 The calculation of the sizing efficiency of a classifier or cyclone can be closely estimated from size analyses of the feed and the products.

4.5 SUB-SIEVE SIZING

Sizing of the particles having size less than 40 microns is known as **Sub-sieve sizing**. The particles at fine sizes are termed as slimes and colloids. The following are the approximate size ranges of different particles:

Particles	Size ranges
Sands	2 mm to 74 microns
Slimes	74 microns to 0.1 microns
Colloids	0.1 micron and 0.001 micron

Table 4.4 Size analysis methods for sub-sieve sizing.

Method	Approximate range, μm
Sedimentation (gravity)	40–1
a Beaker decantation	
b Andresen pipette	
Elutriation	40–5
Sedimentation (centrifugal) Cyclosizer	5–0.05
Microscope (optical)	150–1
Electron microscope	1–0.005
Electrical Resistance method	
Coulter counter	400–0.5

Size analysis methods used for the particles less than 40 microns are shown in Table 4.4.

Sedimentation and elutriation techniques are based on the settling behaviour of the particles of various sizes and the analysis is made by separating the particles into various size fractions. Microscopic sizing is similar to measurement with a yardstick. The particles are sprinkled on a glass slide or mounted in some way on a slide and the size of individual particles is measured under the microscope. Laser beam particle size analysis is of recent origin. The PSM system has been installed in several Mineral Beneficiation plants for continuous measurement of particle size.

In industry, classifiers and hydrocyclones are used for sizing sub-sieve size particles. Fine particles are more difficult to handle and beneficiate. Greater stress is to be given on the process of fine particles not only to recover the values but also to control pollution.

Chapter 5

Screening

Screening is an operation used for the separation of particles according to their sizes. Sieving and screening are distinguished by the fact that sieving is a batch process used almost exclusively for test purposes, whereas screening is a continuous process and is used mainly on an industrial scale. Sieves are manufactured with definite dimensions and standard aperture sizes. Screens can be manufactured with any dimension and any aperture sizes as per the requirement. In industrial screening, the particles of various sizes are fed to the screen surface. The material passing through the screen aperture is called **underflow** (undersize or fines) while the material retained on the screen surface is called **overflow** (oversize or coarse).

5.1 PURPOSE OF SCREENING

Industrial screening is used:

1 To remove oversize material before it is sent to the next unit operation as in closed circuit crushing operations.
2 To remove undersize material before it is sent to the next unit operation which is set to treat material larger than this size.
3 To grade materials into a specific series of sized (finished) products.
4 To prepare a closely sized (the upper and lower size limits are very close to each other) feed to any other unit operation.

Screening is generally used for dry treatment of coarse material. Dry screening can be done down to 10 mesh with reasonable efficiency. Wet screening is usually applied to materials from 10 mesh down to 30 mesh (0.5 mm) but recent developments in the Sieve Bend Screen have made the wet screening possible at the 50 micron size.

5.2 SCREEN SURFACES

Screen surface is the medium containing the apertures for the passage of the undersize material. Several types of screen surfaces are described in Table 5.1.

Table 5.1 Types of screen surfaces.

Type of screen surface	Description	Applications
Parallel rods or **Profile bars**	Rod/bar **Cross sections** Circular, Triangular, Wedge etc.	used for lumpy and coarser size particles
Punched or **perforated plates**	**Openings** Circular, In-line and Staggered openings Square, In-line and Staggered openings Slot-like, In-line and Staggered openings	used for coarser and small sizes slotted openings are sometimes used for fine particles
Woven wires	**Openings** Square Rectangle Triple shute elongated	used for fairly coarse particles used for fine particles used for fine particles

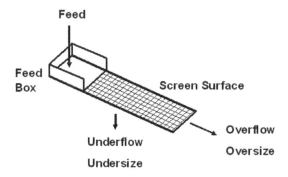

Figure 5.1 Simplified screen.

5.3   SCREEN ACTION

Consider a simplified screen as in Figure 5.1.

The material is fed at one end of the screen. Screening is effected by continuously presenting the material to be sized (the feed) to the screen surface which provides a relative motion with respect to the feed. The screen surface can be fixed or moveable. Agitation of the bed of material must be sufficient to expose all particles to the screen apertures several times during the travel of the material from feed end to the discharge end of the screen. At the same time the screen must act as a transporter for moving retained particles from the feed end to the discharge end. Particles of size more than the aperture size of the screen are retained and smaller particles are passed through the apertures. Both the oversize and undersize particles are collected as overflow and underflow separately.

5.4   TYPES OF SCREENS

Most commonly, screens are used for size separations in conjunction with crushing operations. In the mineral industry, screens are rarely used for separations below 0.2 mm because they have inadequate capacity. However, sieve bends are used for separations as low as 50 μm since these devices give sharper separations than wet classifiers.

Screens are classified as stationary and dynamic screens as shown below:

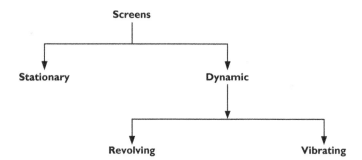

Principal types of Industrial Screens

Stationary Screens

Figure 5.2 Grizzly.

Grizzly
Equally spaced parallel rods or bars
  running in flow direction.
Sloped to allow gravity transport.

Applications
Lumpy or coarse separations.
Scalping before crushing.
Dry separation.

Figure 5.3 Divergator.
(Courtesy Mogensen).

Divergator
Parallel rods running in flow direction.
Fixed at one end.
Gap increases from fixed to free end.
Alternate rods diverge at 5°–6°.

Applications
Separations in the range
  400 to 25 mm size.
Self cleaning and blockage free.
Dry separation.

Figure 5.4 Sieve bend.
(Courtesy Deister
Concentrator, LLC).

Sieve Bend
Stationary curved screen with horizontal
  wedge bars at right angles to slurry
  flow.
Feed slurry enters tangentially.
Imparts centrifugal action.

Applications
Separations in the range of 2 mm to
  45 µm. Wet separation.

Revolving Screens

Figure 5.5 Trommel.
(Courtesy Qingzhou Yuanhua
Machinery Manufacture Co.).

Trommel
Rotating, punched or woven wire.
Slightly inclined cylindrical shell.

Applications
Separations in the range of 10 to 60 mm.
Dry if coarse, wet if fine.
Also used for scrubbing lumpy or coarse.

Vibrating Screens

Figure 5.6 Vibrating grizzly.

(Courtesy Mogensen).

Vibrating Grizzly
Similar to stationary grizzly.
Mechanical or Electrical vibrations.

Applications
Coarse and Dry separations.
Also used as feeders.

Figure 5.7 Vibrating screen.

(Courtesy Robert Cort Ltd, UK).

Vibrating Screen
High speed motion to lift particles.
Mechanical or Electrical vibrations.
Both horizontal and inclined types.

Applications
Separations from 200 mm to
250 μm.
Dry if coarse, wet if fine.

The shaking screen, mounted either horizontally or with a gentle slope, has a slow linear motion essentially in the plane of the screen. Particles slide jerkily and remain in contact with screen surface during screening. Shaking screens may have different aperture surfaces in series so as to prepare a number of different grades. Shaking screens are used for coarser particles down to 12 mm size. These are widely used for coarse coal sizing.

The Reciprocating screen, Gyratory screen, Rotating probability screen, Resonance screen, and Mogensen sizer are some of the other dynamic screens. They are similar to vibrating screens but differ in the type of motion given to the screen deck.

The industrial screens are arranged as single-deck and multi-deck screens. The screen having one screening surface is called a single-deck screen and if a screen has two or more screen surfaces, it is called a multi-deck screen.

Screening is performed either dry or wet. Wet screening is superior, adhering fines are easily washed off, and it avoids the dust problem. But the cost of dewatering and drying the products is high.

The following are some terms used for screens in the mineral industry according to the purpose:

1 **Feed screen:** used to prepare the feed to any unit operation.
2 **Trash Screen:** used to remove the trash material.
3 **Scalping screen:** used to remove small amounts of either oversizes or undersizes.
4 **Dewatering screen:** used to remove water from mixture of solids and water.
5 **Desliming screen:** used to remove slimes from the coarse material.
6 **Medium recovery screen:** used to remove medium solids from coarse material.

Sometimes, as in the case of Iron ores, screening yields sized material of the required grade. Hence screening becomes a beneficiation operation, as practiced at almost all Iron Ore Beneficiation plants in India where beneficiation is done at coarser sizes.

5.5   FACTORS AFFECTING THE RATE OF SCREENING

A number of factors determine the rate at which particles pass through a screen surface and they can be divided into two groups: those related to particle properties and those dependent on the machine and its operation. Some important factors are:

Material factors

1   Bulk density of the material.
2   Size and size distribution of the particles.
3   Size of the particle relative to the aperture.
4   Shape of the particle.
5   Moisture content of the material.

Machine factors

1   Size of the aperture.
2   Shape of the aperture.
3   Size of the screen surface.
4   Percent opening area.
5   Angle of incidence of the particle on the screen surface.
6   Speed at which the particle strikes the screen surface.
7   Thickness of the material on the screen surface.
8   Blinding of the screen surface.
9   Type of screening, i.e., wet or dry screening.
10   Type of motion given to the screen surface.
11   Slope of the screen deck.
12   Mechanical design for supporting and tightening the screen deck.

5.6   SCREEN EFFICIENCY

Screen efficiency (often called the effectiveness of a screen) is a measure of the success of a screen in closely separating oversize and undersize materials. There is no standard method for defining the screen efficiency. Screen efficiency can be calculated based on the amount of material recovered at a given size. In an industrial screening operation, it is to be specified whether the required material is oversize or undersize or both. For the oversize material,

$$\text{Screen efficiency} = \eta = \frac{\text{Weight of actual oversize material present in the feed}}{\text{Weight of overflow material obtained from the screen}}$$

For the undersize material,

$$\text{Screen efficiency} = \eta = \frac{\text{Weight of underflow material obtained from the screen}}{\text{Weight of actual undersize material present in the feed}}$$

The other two ways of representing the screen efficiency is the recovery of oversize material into the screen overflow and the recovery of undersize material into the screen underflow.

The overall efficiency is defined as the product of the recovery of oversize material into the screen overflow and the recovery of undersize material into the screen underflow. The following is the expression used for determining the efficiency of an industrial screen:

$$\eta = \frac{c(f-u)(1-u)(c-f)}{f(c-u)^2(1-f)} \tag{5.6.1}$$

where

f = fraction of oversize material in the feed
c = fraction of oversize material in the overflow obtained from the screen
u = fraction of oversize material in the underflow obtained from the screen

Another expression for efficiency of an industrial screen is given by

$$\eta = \frac{u(u-f)(1-c)(f-c)}{f(u-c)^2(1-f)} \tag{5.6.2}$$

where

f = fraction of undersize material in the feed
c = fraction of undersize material in the overflow obtained from the screen
u = fraction of undersize material in the underflow obtained from the screen

The capacity of an industrial screen is defined as the quantity of material screened per unit time per unit surface area of the screen and is expressed as tons/hr/m². The capacity of screens depends upon (1) the area of the screen surface (2) the size of the screen aperture (3) characteristics of an ore (4) the type of screening mechanism used. Efficiency and capacity are opposite to each other in the sense that capacity can be increased at the expense of efficiency and vice versa. It means that as the capacity increases, efficiency decreases and as the capacity decreases, efficiency increases.

Chapter 6

Liberation

Liberation is the first and the most important step in Mineral Beneficiation. The second step, separation, is impracticable if the first step, liberation, is not accomplished successfully.

Liberation: It can be defined as the freeing or detachment of dissimilar mineral grains. The operation employed to liberate the dissimilar mineral grains is Size reduction or Comminution.

Free particles: If the particles of ore consist of a single mineral, they are termed as Free particles.

Locked particles: If the particles of ore consist of two or more minerals, they are termed as locked particles. If the locked particles contain valuable minerals at considerable quantity, they are termed as middling particles.

Grain size: It is the size of a mineral as it occurs in the ore.

Particle size: It is the size of any particle whether free or locked particle.

Grain and Grain size pertain to uncrushed ore and Particle and Particle size pertain to crushed or ground ore.

Liberation size: It is the size of a mineral particle at which that mineral is completely liberated. It is the size of a free particle of required (valuable) mineral.

Various mineral grains, present in the ore, exist in physical combination with each other. To detach the valuable mineral grains from all other gangue mineral grains, it is essential to reduce the size of the ore particles.

If one mineral species in an ore is to be separated physically from all other species in the ore, all grains of the desired species must be physically detached from all remaining species in the ore. In an ore containing mineral A, B and C (Figure 6.1) if all grains of mineral A (considered as valuable mineral) are to be separated from the ore, all grains of A must be detached from minerals B and C (gangue minerals). When such detachment is complete, mineral A is said to be liberated. However liberation of mineral A does not require liberation of minerals B and C in the ore.

If the physical properties of the adjacent minerals are sufficiently dissimilar, or if the bond between them is notably weaker than either of them, fracture may take place (when it is comminuted) preferentially at the boundary. Comminution results in true freeing or detachment of minerals. Then it is known as **Liberation by detachment** or

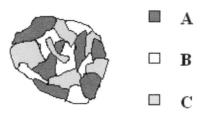

Figure 6.1 A particle of an ore containing A, B and C minerals.

Intergranular liberation. When the run-of-mine ore is reduced in size to grain size of valuable mineral, all the valuable mineral particles exist as free particles, no matter whether remaining particles are free or locked particles of gangue minerals.

Accordingly, the reduced ore consists of:

Free particles of valuable minerals
Free and/or locked particles of gangue minerals

At this condition, particle size = liberation size = grain size

This condition normally does not occur in practice.

If the physical properties of the adjacent minerals are not so dissimilar, comminution to grain size does not result in rupturing the bond between adjacent dissimilar minerals. Fracture occurs across the grains producing locked particles. Hence to detach valuable mineral, the ore is further reduced in size. Thus the reduced ore consists of:

Free particles of valuable mineral
Very few locked particles of valuable and gangue minerals
Free and/or locked particles of gangue minerals

This condition is practically considered as equivalent to freeing. This type of liberation is known as **Liberation by size reduction** or **Transgranular liberation**. Liberation of most of the ores comes under this category.

At this condition, particle size = liberation size < grain size

Let us consider a particle in two dimensions (Figure 6.2). Let ABCD be 8 × 8 cm size particle consists of valuable mineral particles (shaded portion) and gangue mineral particles (white portion) of 2 × 2 cm size each in equal proportion. This particle is reduced by size reduction to a 2 × 2 cm size. If the fracture takes place along the boundary lines as shown in Figure 6.2(a), it yields 16 particles of 2 × 2 cm size each. All 16 particles are free particles out of which 8 particles are of valuable mineral and 8 particles are of gangue mineral. Here all the valuable mineral particles are liberated. This type of liberation is called liberation by detachment.

In this case, **Grain size, Particle size** and **Liberation size** are equal.

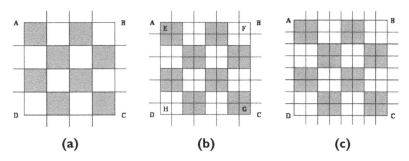

Figure 6.2 Liberation methods.

Suppose if the fracture takes place across the grains as shown in Figure 6.2(b), it yields 25 particles out of which 4 particles E, F, G, H are of 1 × 1 cm size, 12 particles are of 2 × 1 cm size and remaining 9 particles are of 2 × 2 cm size. Here:

E and G are free particles of valuable mineral.
F and H are free particles of gangue mineral.
All others are locked particles.

Hence valuable mineral has been liberated partially to a lesser extent.

If the resulting particles are again crushed and fracture takes place along the boundary lines as shown in Figure 6.2 (c), it yields 64 particles, all are free particles of 1 × 1 cm size out of which 32 are valuable mineral particles and 32 are gangue mineral particles. Here all valuable mineral particles are liberated. This happens when the particles are reduced to 1 × 1 cm size which is less than grain size. 1 × 1 cm is the particle size and also the liberation size.

If the second fracture also does not take place along the boundary lines, it yields still locked particles which need further reduction in size. This reduction is continued till all valuable mineral particles occur as free particles. Then the particle size, hence liberation size, is less than 1 × 1 cm. This happens because the fracture takes place across the mineral grain. This type of liberation is called liberation by size reduction. In this case, **Liberation size (also Particle size)** is less than the **Grain size**.

It is important to note that the particles are reduced by size reduction in both the cases but it matters whether the fracture takes place along the boundaries or across the grains. Locked particles will be produced when the fracture takes place across the grains.

Figure 6.3 shows an example of a typical comminution product wherein black refers valuable mineral and white refers gangue mineral. Only 2 are liberated free valuable mineral particles, 8 are free gangue mineral particles, 6 are locked particles containing valuable and gangue minerals. These locked particles are known as **middling particles**.

Degree of liberation quantitatively referred to as **percent liberation**, of a certain mineral, is defined as the percent of that mineral liberated and occurring as free particles in relation to the total amount of the same mineral present in the ore.

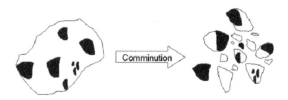

Figure 6.3 Typical comminution product.

For example, the amount of valuable mineral present in one ton of ore is 200 kg. When it is comminuted to a certain size, the amount of valuable mineral existing as free particles is 196 kg. The remaining 4 kg of valuable mineral exists as locked particles with gangue minerals.

$$\text{Then the percent libertion } = \frac{196}{200} \times 100 = 98\%$$

Beneficiation is carried out at this size to separate 98% valuable minerals so that 2% of the valuable minerals are lost. To separate 100% valuable minerals, the ore is to be further reduced in size to get 100% liberation which consumes additional power. If the cost of this additional power is more than the cost of 2% valuable minerals, more than 98% liberation is not desirable.

The beneficiation method to be used depends on liberation size of the ore which in turn depends on type of the ore. Ore types can be conveniently classified as follows:

Massive ores In these ores, reasonable amount of crushing makes the valuables liberated.
Example: Coal, bedded iron ores.
Intergrown ores In these ores, valuables can be freed only partially by crushing and require certain amount of grinding to complete the liberation.
Example: Chrome ore.
Disseminated ores In these ores, valuables are sparely distributed through a waste rock matrix and require fine grinding to liberate the valuables.
Example: Gold ore.

To liberate the valuables, the ore particles are to be reduced in size by the application of the forces. When the forces are applied on the ore particle, fracture takes place depending upon the method of application of the forces.

Chapter 7

Comminution

The operation of applying a force on the particle to break it is called size reduction. **Comminution** is a general term for size reduction that may be applied without regard to the actual breakage mechanism involved.

In any industrial comminution operation, the breakage of any individual particle is occurring simultaneously with that of many other particles. The breakage product of any particle is intimately mixed with those of other particles. Thus an industrial comminution operation can be analyzed only in terms of a distribution of feed particles and product particles. However, each individual particle breaks as a result of the stresses applied to it and it alone.

7.1 FRACTURE

Fracture in the particle occurs as a result of application of a force. When a force is applied on a particle, stress will develop within the particle. When this stress exceeds ultimate stress, the particle will break. Let us consider a particle subjected to two opposing forces by a concentrated load as shown in Figure 7.1.

The principal stress in the z-direction is a compressive stress throughout the particle. The principal stress in the x- and y-directions is a compressive stress adjacent to the load points but a tensile stress within the particle. This tensile stress is lower than compressive stress. As the tensile strength is as little as 1/10 of compressive strength, the fracture occurs primarily because of the tensile stress which results in breakage into a small number of large pieces. Due to the compression adjacent to the loading points, it results a large number of small pieces.

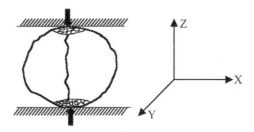

Figure 7.1 Compressive forces.

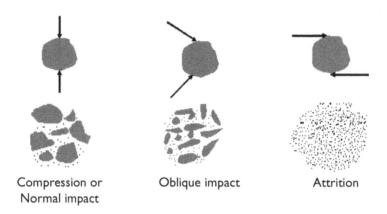

Compression or   Oblique impact   Attrition
Normal impact

Figure 7.2 Mechanism of fracture.

Compressive force can be applied at either a fast or a slow rate. Under the conditions of slow compression, energy applied is just sufficient to load comparatively few regions of the particle to the fracture point and only a few particles result. Their size is comparatively close to the original particle size. Under the conditions of rapid loading such as in high velocity impact, applied energy is well in excess of that required for fracture. Many areas in the particle are overloaded and the result is a comparatively large number of particles with a wide size distribution. Impact causes immediate fracture with no residual stresses.

Attrition or abrasion fracture occurs when a force (shear force) acts parallel to the surface of the particle. Due to insufficient energy applied on the particle, localized stress occurs and a small area is fractured to give very fine particles.

Another type of fracture is chipping. In this chipping, the edges or corners of a particle will break due to the application of oblique forces, generally impact forces, on the particle.

In practice, these events do not occur in isolation. For example, when the particles are crushed by compression as in the case of a crusher, coarse particles will be produced resulting from the induced tensile stress, fine particles will be produced resulting from compressive stress near the points of loading and by attrition due to particle interaction. All these types of forces and fractures, and sizes of the particles after fracture, are shown in Figure 7.2.

It can be summarized that all types of forces exist in any size reduction operation even though individual size reduction units are predominantly designed for application of one type of force.

7.2 LAWS OF COMMINUTION

Laws of comminution are concerned with the relationship between energy input and the size of feed and product particles. Three laws of comminution energy requirements have been put forward by Rittinger, Kick and Bond respectively. None of these three laws is applicable over a wide range of sizes. The Rittinger and Kick laws while tenable in some cases, were never of much use as practical tools. The third law proposed by F.C. Bond (1952) [5] is based on a detailed compilation and study of numerous laboratory and plant comminution data and provides the technician with a reasonably accurate measure of power requirements.

Bond states that the total work useful in breakage that has been applied to a given weight of homogeneous broken material is inversely proportional to the square root of the average size of the product particles, directly proportional to the length of the crack tips formed and directly proportional to the square root of the new surface created. Mathematically, Bond's law is expressed as:

$$W = K_b \left[\frac{1}{\sqrt{P}} - \frac{1}{\sqrt{F}} \right]$$
7.2.1

where
  W = gross energy required, kWhr/short ton
  P = 80% passing size of the product, microns
  F = 80% passing size of the feed, microns
  K_b = Bond's constant

To apply Bond's law, Bond's constant has to be evaluated. Bond's constant is evaluated by defining what is called work index, W_i. It is defined as the gross energy in kWhr/short ton of feed necessary to reduce a very large feed to such a size that 80% of product particles passes 100 microns screen.

Based on this definition, it can be written that:

If $F = \infty$, and $P = 100$ microns, $W = W_i$ kWhr/short ton

On substitution in equation 7.2.1, this gives $W_i = \frac{K_b}{10} \Rightarrow K_b = 10 \, W_i$
Thus, Bond's equation becomes:

$$W = 10 \, W_i \left[\frac{1}{\sqrt{P}} - \frac{1}{\sqrt{F}} \right]$$
7.2.2

The work index includes the friction in the crusher and the power W is gross power. Bond's law is applicable reasonably in the range of conventional rod mill and ball mill grinding.

It is a fact that most of the energy supplied to a comminution machine is absorbed by the machine itself to move various parts of the machine, and only a small fraction of the total energy is available for breaking the material. For example, in a ball mill, less than 1% of the total energy input is used for actual size reduction, the bulk of the energy is utilized in running the mill and in the production of heat.

7.3   OBJECTIVES OF COMMINUTION

The following are some of the objectives of comminution:

1   Reduction of large lumps into small pieces.
2   Production of solids of desired size range.
3   Liberation of valuable minerals from gangue minerals.
4   Preparation of feed material for different beneficiation operations.
5   Increasing the surface area for chemical reaction.
6   Convenience in handling and transportation.

The energy consumed for the comminution operation is high when compared to other operations such as screening, beneficiation, dewatering, conveying etc. in Mineral and Mining Industries. Hence attention needs to be paid to minimize the production of fines (finer than required) which will consume additional power for reducing to fines.

7.4   TYPES OF COMMINUTION OPERATIONS

The run-of-mine ore is quite coarse and cannot be reduced to fine size in one stage. It may require three or more stages. Each stage requires separate equipment. The comminution operations are divided into two broad groups as follows:

1   **Crushing** Crushing is a size reduction operation wherein large lumps are reduced to fragments or smaller particles.
2   **Grinding** Grinding is considered as size reduction of relatively coarse particles to the ultimate fineness.

The machines used for crushing and grinding are entirely different. It is to be noted that the energy required for comminution of unit mass of smaller particles is more than the energy required for unit mass of coarser particles. However, the energy required to reduce coarser particle is more than that of smaller particle. Hence the machines used for crushing (crushers) must be massive and rugged and the machines used for grinding (mills) must be capable of dispersing energy over a large area. In crushers, the breakage forces are applied either by compression or impact whereas in grinding mills shear forces are predominantly applied.

Chapter 8

Crushing

As defined by A.M. Gaudin (1939) [6], crushing is that operation or group of operations in a Mineral Beneficiation plant whose object is to reduce large lumps to fragments, the coarsest particles in the crushed product being 1/20 inch or more in size. The size of coarsest particles is 1/2 inch in many cases.

The crushing action in all crushing machines (crushers) results from forces applied to the particles by some moving part working against a stationary or some other moving part. It is the first stage of size reduction. It has to crush run-of-mine ore which contains large size particles. It requires greater force to be applied on the particles. Crushing is generally a dry operation and is usually performed in two or three stages.

Crushers are designed in such a way that they reduce rock in such a manner that all pieces are less than a stated size. But no crusher has been devised which produces only fragments greater than a specified size. A crusher always produces various sizes of particles with substantial amount of fines. As investigated by Gaudin and Hukki (1944) [7], when the products of crushing are separated into a series of closely sized fractions, the total surface of each fraction will be more or less equal.

The extent of size reduction achieved by any crushing operation is described by the **reduction ratio**. It is defined in a number of ways. Broadly it is defined as the ratio of the maximum size of the particle in the feed to the maximum size of the particle in the product. Two definitions commonly used are termed as average reduction ratio and 80% passing reduction ratio which are defined as follows:

$$\text{Average reduction ratio} = \frac{\text{Average size of the feed particles}}{\text{Average size of the product particles}}$$

$$\text{80\% passing reduction ratio} = \frac{\text{80\% passing size of the feed}}{\text{80\% passing size of the product}}$$

Reduction ratio is a convenient measure for comparing the performance of different crushers.

As the crushing is performed in stages, crushing may be divided into primary, secondary, tertiary and quaternary stages based on the particle size. Correspondingly, the crushers can be classified into five groups according to the size of the product they produce.

1 **Primary Crushers:** Jaw crusher, Gyratory crusher.
2 **Secondary Crushers:** Reduction gyratory, Cone crusher, Rolls crusher.

3 **Tertiary Crushers:** Short-head cone crusher.
4 **Fine Crushers:** Impact crushers.
5 **Special Crushers:** Bradford Breaker, Toothed Roll crusher.

Lumps of run-of-mine ore usually of 1 m size is reduced to 100–200 mm size in heavy duty primary crushers. The usual size of feed to secondary crushers is 600 mm and the product is usually 10–100 mm size. In tertiary crushers, particles of 250 mm size are reduced to 3–25 mm size.

Fine crushers reduce the coarse particle to fine, even to 200 mesh in some cases. Special crushers are designed for specific ores, for example, rotary breaker and toothed rolls crusher for coal and gravity stamps for gold ore milling.

8.1 JAW CRUSHERS

Jaw crushers consist of two jaw plates set at an acute angle, called **angle of nip,** to each other which forms a crushing chamber. One jaw is fixed and kept vertical, the other jaw is a movable or swing jaw and is moved to approach and recede alternately from the fixed jaw. Motion to the swing jaw is transmitted by pitman working on an

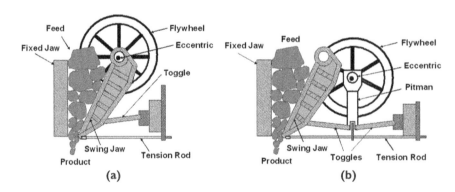

Figure 8.1 (a) Single toggle jaw crusher; (b) Double toggle jaw crusher.

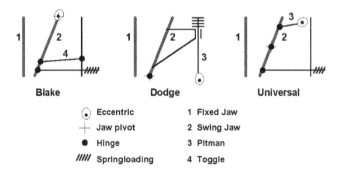

Figure 8.2 Types of jaw crushers.

eccentric and toggles. The material is fed between the jaws and is alternately nipped and crushed. Single and Double Toggle Jaw crushers are shown in Figure 8.1.

Jaw crushers are classified by the method of pivoting the swing jaw (Figure 8.2). If the swing jaw is pivoted at the top, it has a fixed receiving area and variable discharge area and is known as a **Blake crusher**. If the swing jaw is pivoted at the bottom, it has a fixed discharge area and variable receiving area and is known as a **Dodge crusher**. If the swing jaw is pivoted at an intermediate position, it has both variable receiving and discharge areas and is known as a **Universal crusher**.

In a Blake crusher, the distance between the two jaw plates at the feed opening is known as **gape**. The distance between the two jaw plates at the discharge opening is known as **set**. The minimum distance is called **closed set** and maximum distance is called **open set**. The maximum amplitude of swing of the jaw is known as **throw**. In the mining industry, Jaw crushers are used to crush the run-of-mine ore to a size suitable for transportation. The reduction ratio of jaw crushers varies from 4 to 7.

8.2 GYRATORY AND CONE CRUSHERS

Gyratory and Cone crushers are of similar in construction and working. They consist of two vertical truncated conical shells of which the outer hallow conical shell is stationary and the inner solid conical shell is made to gyrate. In a gyratory crusher the inner conical shell is pointing up and outer conical shell is pointing down. The reduction ratio varies from 3 to 10. Figure 8.3 shows the cut section of a gyratory crusher.

Reduction gyratory crusher is the modification of gyratory which has straight or curved heads and concaves and used for secondary crushing. The fine reduction gyratory crusher can also be used for tertiary crushing.

The **cone crusher** is a modified gyratory. Both the outer and inner conical shells are pointing up. **Simons cone crusher** is the most widely used type of cone crusher. It has two forms: **Standard cone crusher** and **Short-head cone crusher**. The outside surface of the standard cone crusher has stepped liners to allow a coarser feed. It yields the product at reduction ratio of 6 to 8. The Short-head cone crusher has a steeper head angle to prevent choking of finer material. The reduction ratio is about 4 to 6.

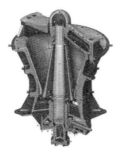

Figure 8.3 Cut section of Fuller-Traylor gyratory grusher.

(Courtesy FLSmidth Minerals).

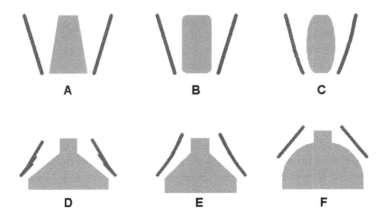

(A) Gyratory; (B) Straight head and concave reduction gyratory; (C) Curved head and concave reduction gyratory; (D) Standard cone; (E) Short-head cone; (F) Gyrasphere

Figure 8.4 Types of crushing chambers.

The Telsmith gyrasphere is another type having a spherical steel head and used for tertiary crushing. Figure 8.4 shows crushing chambers of all types of gyratory and cone crushers.

8.3 ROLL CRUSHERS

Roll Crusher (Figure 8.5) consists of two smooth heavy horizontal cylinders revolving towards each other and the feed material is nipped between the rolls and pulled downward through the rolls by friction. The distinguished feature of a roll crusher is that the material is crushed one time only whilst it is passing through the crushing chamber. Due to this fact, the reduction ratio of a roll crusher varies from 2 to 4, the lowest among all the crushers. Production of fines is minimum. They can handle friable, dry, wet, sticky, frozen, and less abrasive feeds well.

For the selection of size of roll crusher for the reduction of different sizes of feed, two expressions are given by:

$$\cos\frac{n}{2} = \frac{D+s}{D+d}$$ 8.3.1

$$\tan\frac{n}{2} = \mu$$ 8.3.2

where:
    n = angle of nip (Figure 8.6)
    D = diameter of the rolls
    d = diameter of the spherical feed particle
    s = distance between the two rolls (set)
    μ = coefficient of friction between the roll and the particle

Figure 8.5 Roll crusher.

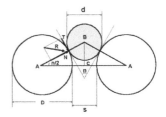

Figure 8.6 Angle of nip of roll crusher.

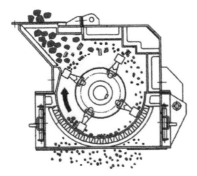

Figure 8.7 Hammer mill.

Smooth-surfaced roll crushers are generally used for fine crushing whereas corrugated or toothed roll crushers are used for coarse crushing of soft materials. **Single toothed roll crusher** and **Double toothed roll crusher** are the two types of toothed roll crushers used for crushing coal.

8.4   IMPACT CRUSHERS

Impact crushers reduce the particles by impact forces applied through sharp blows of fixed or free swinging hammers revolving about central rotor at high speed to the free falling particles against stationary surfaces. They are used for relatively soft, friable, and sticky ores such as phosphates, limestone, clay, graphite and coal. Hammer mill (Figure 8.7) is one type of impact crusher.

8.5   BRADFORD BREAKER

Bradford breaker is a typical machine resembling cylindrical trommel screen in operation. It consists of a slightly inclined cylindrical chamber with a perforated wall (Figure 8.8) and rotates about its axis at a low rpm. It is extensively used for primary crushing of run-of-mine coal.

Figure 8.8 Bradford breaker.

(Courtesy Pennsylvania Crusher Corporation).

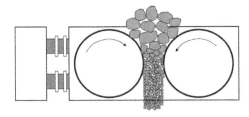

Figure 8.9 High pressure grinding rolls (HPGR).

The coal particles fed into the cylinder are lifted by longitudinal lifters within the cylinder and broken by falling and striking the coal below as the cylinder is rotated. The particles when broken to the required size quickly fall through the perforations thus produce few fines. Unbroken particles are discharged from the end of the breaker.

8.6   HIGH PRESSURE GRINDING ROLLS (HPGR)

In a roll crusher, the force of compression and friction makes the particles to crush. In High Pressure Grinding Rolls (Figure 8.9), rolls are subjected to high pressure so that comminution takes place by compressive forces as well as by inter-particle breakage. The force applied to the crushing zone is controlled by a hydro-pneumatic springs. As the product size from HPGR is fine, it can replace the conventional secondary and tertiary crushers.

8.7   CRUSHING OPERATION

Crushers are usually operated dry. When the material fed to the crusher is at a slow rate, the individual particles are crushed freely. The crushed product is quickly removed from the crushing zone. This type of crushing, known as free crushing, avoids the production of excessive fines by limiting the number of contacts.

When the material fed to the crusher is at a high rate, the crusher is choked and it prevents the complete discharge of crushed product. This results in crushing between the ore particle and the crushing surface as well as between the ore particles. This type of operation increases the amount of fines produced. This type of choked feeding is preferred in some cases as it reduces the reduction stages.

Usually crushing is performed for any ore in two or three stages depending on the size of the feed particles and the size of the product particles required.

8.8   OPEN CIRCUIT AND CLOSED CIRCUIT OPERATIONS

Usually each stage of size reduction is followed by a screen which forms a circuit. Crushing may be conducted in open or closed circuit. Figure 8.10 shows the typical open and closed circuit operations:

In an open circuit crushing operation, the feed material is reduced by one crusher. The product of this crusher is screened and only oversize material is crushed by another crusher of small size as the **throughput** (the quantity of material crushed in a given time) is less. The crushed product from the second crusher and undersize material from the screen together form the final product.

In a closed circuit crushing operation, the oversize material from the screen is fed back to the same crusher. The quantity of the oversize material fed back to the crusher is called as **circulating load.** The undersize material from the screen is the required final product. Initially the quantity of final product produced is less than the quantity of the feed material. As the operation proceeds further the quantity of the final product gradually increases and will be equal to the quantity of the feed material after some time. After attaining this equilibrium condition, the quantity of the final product is always equal to the quantity of the feed material and the circulating load is constant. The circulating load is expressed as a percentage of the quantity of feed material. It is to be noted that the throughput of the crusher is more than that of the crusher used in open circuit crushing operation and equal to the quantity of feed material plus the quantity of oversize material fed back from the screen.

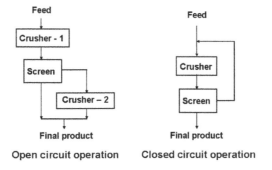

Figure 8.10 Crushing circuits.

It is obvious that two size reduction machines are used in an open circuit operation whereas only one machine is employed in a closed circuit operation to get the final product. The objectives of employing a closed circuit operation are to minimize the production of fines and to reduce the energy consumption by avoiding size reduction of already reduced particles to the required size.

Chapter 9

Grinding

Grinding is the last stage of the comminution process. The particles are reduced from a maximum upper feed size of 3/8 inch, to some upper limiting product size ranging between 35 mesh and 200 mesh (420 microns and 74 microns). Grinding is performed in rotating steel vessels knows as **tumbling mills** or **grinding mills**. A grinding mill consists of a horizontal rotating steel shell supported by end bearings on which hallow trunnions revolve. Loose crushing bodies, known as **grinding medium**, are placed inside the shell. Either steel balls/rods or pebbles are used as grinding medium. They are free to move inside the rotating shell making the particles break by repetitive blows and by rolling and sliding one over the other. Attrition, or shearing, forces which result from the application of forces by rolling and sliding bodies tend to produce more fine particles than impact forces applied on particles by repetitive blows.

Grinding mills can be operated wet or dry, batch-wise or continuously. The equipment is robust and the loose grinding medium can usually be added without stopping the mill. On the other hand, grinding mills are relatively high in power consumption and require expensive foundations. Grinding mills are normally loaded to approximately 50 percent of its volume with grinding medium.

The interior of tumbling mill is lined by replaceable liners usually made of alloy steel but sometimes of rubber. Some types of liners are smooth, shiplap, wave, wedge bar, rib, stepped, osborn, lorain, etc. Smooth liners favour abrasion resulting in fine grinding but high metal wear. Liners other than smooth are designed to help in lifting the ball load as the mill is revolved and sometimes to minimize the slip between the layers of balls. Liners protect the mill body from wear and damage.

Usually the material is fed at one end of the mill and discharged at the other end. In dry grinding mill, the feed is by vibrating feeder. Three types of feeders (Spout, Drum and Scoop feeders) are in use to feed the material to the wet grinding mills. In a **spout feeder**, the material is fed by gravity through the spout. In **drum feeder**, the entire mill feed enters the drum and an internal spiral carries it and fed to the mill. Grinding balls are conveniently added through this feeder during operation. In case of **scoop feeder**, material is fed to the drum and the scoop picks it up and fed to the mill.

Grinding mills are classified as **Ball mill, Rod mill, Tube** or **Pebble mill,** and **Autogenous mill** on the basis of grinding medium and shell proportions. In ball mills, the grinding medium is steel balls; in rod mills, steel rods; in tube or pebble mills, pebbles of hard rock or other nonmetallic material; in autogenous mills, coarse ore particles.

9.1 BALL MILL

Ball mill uses steel or iron balls as grinding medium. Ball mills usually have a length to diameter ratio of 1.5 to 1.0. According to the shape of the mill, the ball mills are classified as cylindroconical and cylindrical mills (Figure 9.1).

As the mill rotates, the balls are lifted to certain height and then dropped. Grinding of ore particles takes place due to simple rolling of one ball over the other (cascading) and by the free fall of balls (cataracting). Cascading leads to fine grinding whereas cataracting leads to coarse grinding. Figure 9.2 shows the motion of the charge in the ball mill.

As the speed of the mill increases, the balls are lifted higher and a stage is reached where the balls are carried around the shell and never allowed to fall. That means centrifuging occurs. The balls will rotate as if they are part of the shell. The speed at which centrifuging occurs is known as **critical speed**. An expression for the critical speed is given by:

$$\text{Critical speed} = N_c = \frac{42.3}{\sqrt{(D-d)}} \text{ revolutions/minute} \qquad 9.1.1$$

where D and d are the diameter of the Mill and the ball in metres.

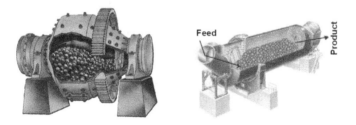

Figure 9.1 (a) Cylindroconical ball mill; (b) Cylindrical ball mill.
(Courtesy Metso Minerals); (Courtesy www.mine-engineer.com).

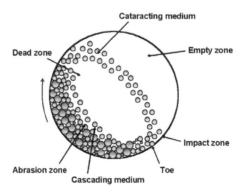

Figure 9.2 Motion of the charge in a ball mill.

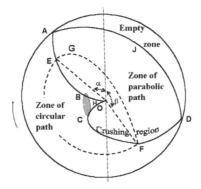

Figure 9.3 Zones in a ball mill.

Davis [8] has given an exhaustive mathematical analysis of the action in a ball mill and explained various zones in a ball mill (Figure 9.3).

As shown in Figure 9.3, FE is the circular path and EGF is the parabolic path of a ball placed at point F or E. The locus of the point E where the ball changes from circular path to parabolic path for all positions of the balls from center of the mill to the periphery is OBEA. The locus of the point F, the end of the parabolic path for all positions of the balls is DFCO. Davis has shown that the arcs CO and BO correspond to unstable equilibrium. The zone BCH is a dead zone where there is no effective motion, hence no grinding takes place. Thus the inside of the ball mill consists of four zones:

1   An empty zone where no balls occupy this zone during operation.
2   A dead zone where no grinding takes place.
3   A zone of circular path where balls roll on each other and grinding takes place by slippage between ball layers before they are lifted.
4   A zone of parabolic path where the balls spread out and fall down.

When the speed of the mill exceeds the critical speed, all these zones disappear, the balls will centrifuge and no grinding takes place anywhere inside the mill. Hence the mill should be operated at a speed below the critical speed. The usual range is 60–80% of critical speed.

Balls vary in size from 1 to 6 inches. The largest balls are used for coarser grinding. Initially, at the start of a ball mill, balls of various sizes, known as **seasoned charge**, are introduced. As the balls wear out gradually, only the largest balls are added as make up media.

9.2   ROD MILL

A rod mill uses rods as a grinding medium. Length to diameter ratio is between 1.5 and 2.5. Rods are a few centimeters shorter than the length of the mill to avoid

Figure 9.4 Grinding action of rods.

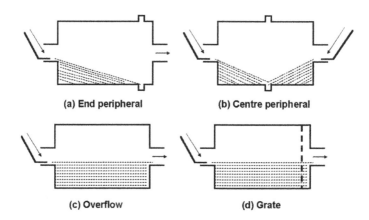

(a) End peripheral   (b) Centre peripheral

(c) Overflow   (d) Grate

Figure 9.5 Types of rod mills according to method of discharge.

any jamming of rods in the mill. The rods are kept apart by the coarsest particles. The grinding action results from line contact of the rods on the ore particles and is exerted preferentially on the coarsest particles. Smaller and fine particles do not grind till the coarsest particle is reduced in size. This can clearly be observed in Figure 9.4. Thus the rod mill produces a more closely sized product with little oversize or slimes. Hence the rod mills may be considered as coarse grinding machines.

If the material inside the mill is made to discharge through the periphery of the mill shell, it passes through quickly and less over-grinding takes place. Rod mills are classed according to the method of the discharge of ground product as **Peripheral, Overflow** and **Grate discharge mills** (Figure 9.5).

There are two types in peripheral discharge mills. In the end peripheral discharge mills, the material is fed at one end of the mill and the ground product is discharged from the other end by means of several peripheral apertures into a close-fitting circumferential chute. In centre peripheral discharge mills, the material is fed at both ends of the mill and the ground product is discharged through a circumferential port at the centre of the shell. The material is fed at one end and the product is discharged through the other end by overflow in overflow discharge mill. In a grate discharge mill, discharge grates are fitted through which pulp flows freely and lifted up to the level of the discharge trunnion.

Rod mills are normally run at 50–65% of the critical speed so that the rods cascade rather than cataract.

9.3 TUBE OR PEBBLE MILL

A tube mill is a grinding mill where length to diameter ratio is 3 to 5. It is also known as a pebble mill as it uses ceramic pebbles made of flint or porcelain as the grinding media. As pebbles are fragile, pebble mills are smaller in diameter. Because of low specific gravity of pebbles, large quantity of pebbles has to be used to attain required grinding for equivalent duty as compared with ball mills. Pebble mills are used when iron contamination in the product is highly objectionable such as in the manufacture of paints, pigments, cosmetics, etc.

Tube mills are sometimes divided into several longitudinal compartments, each having a different charge composition. The charge can be steel balls, rods or pebbles and they are often used dry to grind cement clinker, gypsum and phosphate.

9.4 AUTOGENOUS MILL

Autogenous mill use coarse ore particles as grinding medium. Grinding is achieved by the action of ore particles on each other when particles of ore of different sizes are rotated together in a tumbling mill. The mill is of very large in diameter. The larger particles in the feed must be sufficient in size and number to break the smaller particles as fast as they themselves are broken down in the mill. Autogenous grinding greatly reduces metal wear and also reduces the number of crushing and grinding stages as compared to conventional size-reduction operations, thus it offers the possibilities of large savings in capital and operating costs. However, it is practical for only a limited number of ore types.

Autogenous grinding differs from grinding with metal balls or rods in that the breakage is much more confined to zones of weakness in the rock, such as crystal surfaces and fine cracks. An ore with a pronounced natural grain size, or crystal size, may be ground autogenously to that size with comparative ease, but is ground finer only with difficulty.

Semi-Autogenous mill (SAG mill) uses a combination of the ore and a reduced charge of balls or rods as a grinding medium to overcome the difficulties encountered in autogneous grinding.

9.5 WET AND DRY GRINDING

The grinding may be wet or dry depending on the subsequent process and the nature of the product. Wet grinding is generally used in mineral beneficiation plants as subsequent operations for most of the ores are carried out wet. Wet grinding is usually carried out with 20–35% water by weight. The chief advantages of wet grinding are increased capacity (as much as 15%) for a given size of equipment and less power consumption per ton of the product. Low power consumption is due to the penetration of water into the cracks of the particles which reduces the bond strength at the crack tip.

Dry grinding is used whenever physical or chemical changes in the material occur if water is added. It causes less wear on the liners and grinding media. Dry grinding mills are often employed to produce an extremely fine product. This arises from the high settling speed of solids suspended in air as compared with solids suspended in water.

9.6 GRINDING CIRCUITS

Mesh of grind (m.o.g.) is the term used to designate the size of the grounded product in terms of the percentage of the material passing through a given mesh. In grinding, there are always some particles which may repeatedly be reduced to fine size whereas some other particles may not be reduced. The primary objective of a grinding mill is to reduce all particles to the stated size i.e., mesh of grind. When a grinding mill is fed with a material, it should be at a rate calculated to produce the correct product in one pass in which case it is known as Open circuit grinding (Figure 9.6).

There is no control on product size distribution in the open circuit grinding.

In closed circuit grinding (Figure 9.7), the material of the required size is removed by a classifier from the ground product to send to the subsequent operation and oversize is returned to the same grinding mill.

Special feature of closed circuit grinding is the removal of the product from the grinding mill as soon as the material is grounded to the required size so that overgrinding (grinding to the size finer than required) is avoided.

In case of a rod mill, rods exert sizing action, hence the use of closed circuit grinding is not necessary.

In closed circuit grinding, the amount of solids by weight fed back to the grinding mill is called **circulating load** and its weight is expressed as a percentage of the weight of new feed.

$$\text{Percent circulating load} = \frac{U}{NF} \times 100 \qquad\qquad 9.6.1$$

where U is the circulating load

NF is the amount of new feed solids fed to the mill

The grinding mills are generally operated at circulating loads of 200–500% in order to have the grinding to the required size.

Figure 9.6 Open circuit grinding.

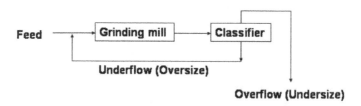

Figure 9.7 Closed circuit grinding.

Chapter 10

Separation operations

The word **separation** is to be used carefully as it conveys different meanings in different situations. The three following types of separation operations are usually involved in Mineral Beneficiation Plants:

10.1 SEPARATION OF ORE PARTICLES ACCORDING TO THEIR SIZE

Operations: Screening
Sizing Classification
Centrifugal Sizing

10.2 SEPARATION OF ORE PARTICLES ACCORDING TO A PROPERTY WHERE VALUABLE MINERAL PARTICLES ARE DIFFERENT FROM GANGUE MINERAL PARTICLES IN THAT PROPERTY

Operations: Ore Sorting by hand and by mechanical means
Sorting Classification
Gravity Separation or Concentration

a Heavy Liquid Separation
b Heavy Medium Separation
c Jigging
d Spiraling
e Tabling

Centrifugal Separation
Flotation
Magnetic Separation
Electrical Separation

10.3 SEPARATION OF ORE PARTICLES FROM THE MIXTURE OF SOLIDS AND FLUIDS

Operations: Dewatering by screening
Thickening
Filtration
Drying

The ore particles separated by the first type of operations are fed to the second type of operations which are the actual beneficiation operations.

Separation of ore particles according to their size by the first type of operations is needed because each beneficiation operation in the second type is suitable for a particular size range of particles.

If the beneficiation is done by the medium of air, it is called **pneumatic** or **dry** beneficiation. The products are directly sent for further processing.

If the beneficiation is done by the medium of water, it is called **wet** beneficiation. The products, which contain water, are sent to the third type of operations to remove the water and to make them suitable for further processing.

Among the above three types of separation operation, Classification, Gravity separation and Thickening are based on the settling of solid particles in a fluid medium. Even though the conditions maintained in each operation are different, the basic principles of settling are the same.

An understanding of the density and other terminology concerned with the mixture of water and solid particles is important in the study of the principles of settling.

Chapter 11

Density

Density is important next to the size of the particle. Density of the particle is defined as the mass of the particle per unit volume. The ratio of the density of the particle to the density of water is defined as specific gravity.

In all beneficiation operations, particularly in Gravity Concentration operations, density, together with the size and the shape of the particles, has an important role. As the ore contains different minerals, the density of an ore varies depending on the minerals it contains. Before the ore is to be beneficiated, the density of the ore and the density of the different minerals present in the ore are to be determined.

A bulk solid (bulk material) is a combination of particles and space. For a bulk material, the average particle density can be determined by dividing the mass of the material (solids) by the true volume occupied by the particles (not including the voids). This can be determined by using a density bottle.

The stepwise procedure for determination of density of an ore is as follows:

1   Wash, dry and weigh the density bottle with stopper. Let this weight be w_1.
2   Thoroughly dry the ore sample.
3   Add 5–10 grams of ore sample to the bottle and reweigh. Let this weight be w_2.
4   Now fill the bottle with a liquid of known density. The liquid used should not react with the ore.
5   Insert the stopper, allow the liquid to fall out of the bottle, wipe off excess liquid and weigh the bottle. Let this weight be w_3.
6   Remove ore and liquid from the bottle and fill the bottle with liquid alone and repeat step 5. Let this weight be w_4.

$w_2 - w_1$ is the weight of the ore sample.
$w_4 - w_1$ is the weight of the liquid occupying whole volume of the bottle.
$w_3 - w_2$ is the weight of the liquid having the volume equal to the volume of density bottle less volume of ore sample taken.
$(w_4 - w_1) - (w_3 - w_2)$ is the weight of the liquid of volume equal to that of the ore sample.

If ρ_l is the density of the liquid:

$$\text{Density of the ore sample } = \frac{w_2 - w_1}{(w_4 - w_1) - (w_3 - w_2)} \times \rho_l \qquad 11.1$$

Most of the mineral beneficiation operations are wet. Water is added to the ore particles to aid beneficiation. The mixture of water and solid particles is known as **Pulp**.

Other terms commonly used are:

Suspension: When the solid particles are held up in the water, the pulp is called suspension. In other words, in suspension, the solid particles are well dispersed throughout.

Slurry: A mixture of fine solids (slimes) and water.

Sludge: Thick pulp i.e., pulp with less quantity of water.

Pulp or slurry density is most easily measured in terms of the weight of the slurry per unit volume (gm/cm^3 or kg/m^3). A sample of slurry taken in container of known volume is weighed to give slurry density directly. Marcy Scale available in the market gives direct reading for the density of the slurry and % solids in the slurry.

The composition of a slurry is often represented as the fraction (or percent) of solids by weight. It is determined by sampling the slurry, weighing, drying and reweighing.

$$C_w = \text{fraction of solids by weight} = \frac{\text{Weight of the particles}}{\text{Weight of the slurry}}$$

$$C_v = \text{fraction of solids by volume} = \frac{\text{Volume of the particles}}{\text{Volume of the slurry}}$$

Knowing the densities of the slurry (ρ_{sl}), water (ρ_w) and dry solids (ρ_p), the fraction of solids (C_w) by weight can be calculated. Since the total volume of the slurry is equal to the volume of the solids plus the volume of the water, then for unit volume of the slurry:

$$\frac{C_w}{\rho_p} + \frac{1-C_w}{\rho_w} = \frac{1}{\rho_{sl}} \qquad\qquad 11.2$$

$$\Rightarrow \quad C_w = \frac{\rho_p(\rho_{sl}-1)}{\rho_{sl}(\rho_p-1)} \quad [\because \rho_w = 1 \text{ gm/cm}^3] \qquad\qquad 11.3$$

Similarly, the total weight of the slurry is equal to the weight of the solids plus the weight of the water, then for unit weight of the slurry:

$$C_v\,\rho_p + (1 - C_v)\,\rho_w = \rho_{sl} \qquad\qquad 11.4$$

$$\Rightarrow \quad C_v = \frac{(\rho_{sl}-1)}{\rho_p-1} \quad [\because \rho_w = 1 \text{ gm/cm}^3] \qquad\qquad 11.5$$

From the grinding stage onwards, most mineral beneficiation operations are carried out on slurry streams. The slurry is transported through the circuit via pumps and pipelines. The water acts as a transportation medium. The volume of the slurry flowing through the circuit will affect the residence time in unit operations. Volumetric flow rate can be measured by diverting the stream of slurry into a suitable container for a measured period of time. The ratio of the volume of the slurry collected to the time taken to collect the slurry gives the flow rate of the slurry. This volumetric flow rate is important in calculating retention time of the slurry in any operation. For instance, if 180 m³/hr of the slurry is fed to a flotation conditioning tank of a volume of 30 m³ then, on an average, the retention time of particles in the tank will be:

$$\text{Retention time} = \frac{\text{Tank Volume}}{\text{Flow rate}} = \frac{30}{180} = \frac{1}{6}\text{hr} = 10 \text{ minutes}$$

That means, any part of the slurry takes 10 minutes from the time it enters the tank to the time it leaves the tank.

Dilution ratio is the ratio of the weight of the water to the weight of the solids in the slurry.

$$\text{Dilution ration} = \frac{1 - C_w}{C_w} \qquad\qquad 11.6$$

Dilution ratio is particularly important as the product of dilution ratio and weight of the solids in the slurry is equal to the weight of the water in the slurry.

When it is required to prepare a liquid of definite density by mixing two miscible liquids of known densities as in case of float and sink analysis, the equation is written as:

$$C_v \rho_1 + (1 - C_v) \rho_2 = \rho_{12} \qquad\qquad 11.7$$

where
ρ_1 = density of liquid 1
ρ_2 = density of liquid 2
ρ_{12} = density of the resultant liquid after mixing two liquids
C_v = fraction of liquid 1 by volume

$$\Rightarrow \quad C_v = \frac{\rho_{12} - \rho_2}{\rho_1 - \rho_2} \qquad\qquad 11.8$$

If C_w is fraction of liquid 1 by weight, the equation is:

$$\frac{C_w}{\rho_1} + \frac{1 - C_w}{\rho_2} = \frac{1}{\rho_{12}} \qquad\qquad 11.9$$

$$\Rightarrow \quad C_w = \frac{\rho_1(\rho_{12} - \rho_2)}{\rho_{12}(\rho_1 - \rho_2)} \qquad\qquad 11.10$$

Chapter 12

Settling of solids in fluids

One of the most effective techniques used for the separation of fine solid particles is sedimentation. Sedimentation is the act of the settling of solid particles in a fluid medium under the force of gravity. A few observations in everyday life indicate that there are natural forces which can, under control, be used in the mineral beneficiation field to separate one mineral from another or to separate a solid from a fluid. The following are some examples:

1   If a stone and a feather are dropped from the same height in air, the stone lands first – due to the difference in shape causing different resistances.
2   Sawdust floats and sand sinks in a pail of water – due to the difference in the specific gravities of the three materials.
3   The texture of a river bottom becomes finer as one approaches the river mouth.

All the above examples are uncontrolled manifestations of forces of gravity and fluid resistance. If controlled, these forces can be used in mineral beneficiation to effect the required separation.

12.1   PRINCIPLES OF SETTLING

Consider a single homogeneous spherical particle of diameter 'd' and density 'ρ_p' falling under gravity in a viscous fluid of density 'ρ_f' and viscosity 'μ_f'. Let the particle is falling in a stationary fluid extending in all directions to infinity in a uniform field of force.

There are three forces acting on a particle:

1   **Gravity force,** $m_p g$, product of the mass of the particle (m_p) and the acceleration due to gravity (g), acts downwards.
2   **Buoyant force,** $m_f g$, (by Archimedes' principle) product of the mass of the fluid displaced by the particle (m_f) and the acceleration due to gravity (g), which acts parallel and opposite to the gravity force.
3   **Drag force, R,**   (resistance to the motion), which acts on the surface of the particle and is parallel, and opposite, to the gravity force. This force increases with velocity.

According to Newton's second law of motion, the equation of motion of the particle is:

$$m_p g - m_f g - R = m_p \frac{dv}{dt} \qquad 12.1.1$$

where v is the velocity of the particle and $\frac{dv}{dt}$ is the acceleration of the particle.

If the drag force, or resistance force, becomes equal in magnitude and opposite in direction to the resultant of the other two forces (gravity and buoyancy) acting on a particle in a fluid, the acceleration of the particle will be nil and the velocity is constant.

This velocity is the maximum velocity attained by the particle. It is known as **maximum velocity** or **terminal velocity** (v_m). Once the particle attains this velocity, it will fall with the same velocity thereafter.

When the acceleration is zero, the particle attains the terminal velocity. Hence when $\frac{dv}{dt} = 0$ equation **12.1.1** becomes:

$$R = g(m_p - m_f) \qquad 12.1.2$$

$$\Rightarrow \quad R = g\left(\frac{\pi}{6}d^3\rho_p - \frac{\pi}{6}d^2\rho_f\right) \qquad \text{as the particle is assumed as sphere}$$

$$\Rightarrow \quad R = \frac{\pi}{6}gd^3(\rho_p - \rho_f) \qquad 12.1.3$$

The nature of the resistance (or drag) depends on the velocity of descent. At low velocities, motion is smooth because the layer of fluid in contact with the body moves with it, while the fluid, a short distance away, is motionless. Between these two positions is a zone of intense shear in the fluid all around the descending particle. Hence the resistance to the motion is due to the shear forces or viscosity of the fluid and is called **viscous resistance.**

Stokes [9], an eminent English Physicist, assumed that the resistance offered by the fluid is due to viscous resistance and deduced as $3\pi d\mu_f v$. Then equation **12.1.3** becomes, after replacing v by v_m, terminal velocity of the particle:

$$3\pi d\mu_f v_m = \frac{\pi}{6}gd^3(\rho_p - \rho_f) \qquad 12.1.4$$

$$\text{On computation, } v_m = \frac{d^2 g(\rho_p - \rho_f)}{18\mu_f} \qquad 12.1.5$$

The expression **12.1.5** is applicable for fine particles of less than 50 microns in size, and is also applicable, with small deviations, up to 100 microns.

As the size of the particle increases, settling velocity increases. At high velocities, the main resistance is due to the displacement of fluid by the particle and is known as **turbulent resistance.** In this case, the viscous resistance is relatively small.

Newton [10] assumed that the resistance is entirely due to turbulent resistance and deduced as $0.055\pi d^2 v^2 \rho_f$. Then equation **12.1.3** becomes, after replacing v by v_m

$$0.055\pi d^2 (v_m)^2 \rho_f = \frac{\pi}{6} g d^3 (\rho_p - \rho_f)$$

12.1.6

On computation, $v_m = \sqrt{\dfrac{3gd(\rho_p - \rho_f)}{\rho_f}}$

12.1.7

The expression 12.1.7 is applicable for the particles of size more than 2 mm (2000 microns).

Whether viscous or turbulent resistance predominates, the acceleration of the particle in a fluid rapidly decreases and terminal velocity is quickly reached.

The terminal velocity of a spherical particle is a function of size and specific gravity (density) of the particle. If two particles have the same specific gravity, then the larger diameter particle has higher terminal velocity and if two particles have the same diameter, then the heavier particle has higher terminal velocity. The velocity of an irregularly shaped particle with which it is settling in a fluid medium also depends on its shape. As almost all natural particles are irregular in shape, one can state that:

The coarser, heavier and rounder particles settle faster than the finer, lighter and more angular particles.

12.1.1 Free settling

In a large volume of fluid, the particle settles by its own specific gravity, size and shape and uninfluenced by the surrounding particles as the particles are not crowded. Such a settling process is called **Free settling**. Free settling predominates in well dispersed pulps where the percent of solids by weight is less than 10. Figure 12.1 (a) shows how the particles of different sizes and two specific gravities settle under free settling conditions.

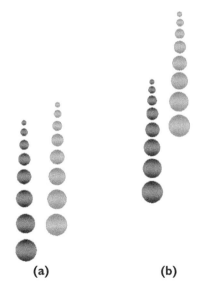

(a)   (b)

Figure 12.1 (a) Free settling; (b) Hindered settling.

12.1.2 Hindered settling

When the particles settle in a relatively small volume of fluid, they are crowded in the pulp and are very close to each other. As a result, the settling of a particle is influenced by surrounding particles. Such a settling process is called **Hindered settling**. In this type of settling, particles collide with each other during their settling and this collision affects their settling velocities. Thus lower settling velocities are encountered.

Hindered settling predominates when the percent of solids by weight is more than 15. Figure 12.1 (b) shows how the same particles, as considered in free settling, settle under hindered settling conditions. By comparing (a) and (b) in Figure 12.1, it is evident that the heavier (or lighter) particles can be separated when they settle by hindered settling. This is possible because hindered settling reduces the effect of size and increases the effect of specific gravity.

12.1.3 Equal settling particles

Particles are said to be equal settling if they have the same terminal velocities in the same fluid and in the same field of force.

12.1.4 Settling ratio

Settling ratio is the ratio of the sizes of two particles of different specific gravities that fall at equal rates.

Under free settling conditions, settling ratio is known as free settling ratio and can be obtained by equating the terminal velocities of lighter and heavier particles of different sizes. Let d_1 and d_2 be the diameters of lighter and heavier particles and ρ_{p1} and ρ_{p2} be the densities of lighter and heavier particles. When the terminal velocities of these two particles are the same and the particles are fine, obeying Stokes' law of settling, the equation for the terminal settling velocity (equation 12.1.5) can be written as:

$$v_m = \frac{d_1^2 g(\rho_{p1} - \rho_f)}{18\mu_f} = \frac{d_2^2 g(\rho_{p2} - \rho_f)}{18\mu_f} \qquad 12.1.8$$

$$\Rightarrow \quad \text{Free settling ratio} = \frac{d_1}{d_2} = \left(\frac{\rho_{p2} - \rho_f}{\rho_{p1} - \rho_f}\right)^{1/2} \qquad 12.1.9$$

When the terminal velocities of the two particles are the same and the particles are coarse, obeying Newton's law of settling, equation 12.1.7 can be written as:

$$v_m = \sqrt{\frac{3gd_1(\rho_{p1} - \rho_f)}{\rho_f}} = \sqrt{\frac{3gd_2(\rho_{p2} - \rho_f)}{\rho_f}} \qquad 12.1.10$$

$$\Rightarrow \quad \text{Free settling ratio} = \frac{d_1}{d_2} = \frac{\rho_{p2} - \rho_f}{\rho_{p1} - \rho_f} \qquad 12.1.11$$

The general expression for a free-settling ratio can be deduced from equations 12.1.9 and 12.1.11 as:

$$\text{Free settling ratio} = \frac{d_1}{d_2} = \left(\frac{\rho_{p2} - \rho_f}{\rho_{p1} - \rho_f}\right)^n \qquad \qquad 12.1.12$$

where $n = 0.5$ for fine particles obeying Stokes' law and $n = 1$ for coarse particles obeying Newton's law. The value of n lies in the range 0.5–1 for particles in the intermediate size range of 100–2000 microns.

Consider a mixture of galena (density 7.5 gm/cc) and quartz (density 2.65 gm/cc) particles settling in water. For fine particles, obeying Stoke's law, the free settling ratio is:

$$\sqrt{\frac{7.5 - 1}{2.65 - 1}} = 1.99$$

i.e., a fine galena particle will settle at the same rate as fine quartz particle of diameter 1.99 times larger than galena particle. Figure 12.2(a) shows the settling of fine particles.

For coarse particles, obeying Newton's law, the free settling ratio is:

$$\frac{7.5 - 1}{2.65 - 1} = 3.94$$

i.e., a coarse galena particle will settle at the same rate as coarse quartz particle of diameter 3.94 times larger than galena particle. Figure 12.2(b) shows the settling of coarser particles.

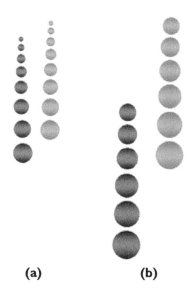

(a)   **(b)**

Figure 12.2 Free settling of (a) Fine particles; (b) Coarse particles.

Therefore the free settling ratio for coarse particles is larger than for fine particles. This means that density difference between the particles has more effect at coarser size ranges when they settle.

As the percentage of solids in the pulp increases, the effect of particle crowding becomes more and the individual particles tend to interfere with each other and therefore the velocity of motion or rate of settling of each individual particle will be considerably less than that for the free settling conditions. The system begins to behave as a heavy liquid whose density is that of the pulp rather than that of carrier liquid. Now the hindered settling conditions prevail. It must be noted that each particle is in fact settling through a suspension of other particles in the liquid rather than through the simple liquid itself.

The effective density and viscosity of a concentrated suspension are much larger than those of a clear liquid. The settling medium therefore offers high resistance and this resistance to fall is mainly due to the turbulence created. Hence Newton's law can be used to determine the approximate terminal velocity of the particles by replacing ρ_f by ρ_{sl} the density of the slurry or pulp.

$$v_m = \sqrt{\frac{3gd(\rho_p - \rho_{sl})}{\rho_{sl}}} \qquad\qquad 12.1.13$$

The lower the density of the particle, the more marked the effect of reduction of the effective density, $(\rho_p - \rho_{sl})$, and the greater the reduction in falling velocity. Similarly, the larger the particle, the greater the reduction in falling rate as the pulp density increases.

Hindered settling ratio can be obtained by equating the terminal velocities of lighter and heavier particles of different sizes (equation 12.1.13).

$$v_m = \sqrt{\frac{3gd_1(\rho_{p1} - \rho_{sl})}{\rho_{sl}}} = \sqrt{\frac{3gd_2(\rho_{p2} - \rho_{sl})}{\rho_{sl}}} \qquad\qquad 12.1.14$$

$$\Rightarrow \quad \text{Hindered settling ratio} = \frac{d_1}{d_2} = \frac{\rho_{p2} - \rho_{sl}}{\rho_{p1} - \rho_{sl}} \qquad\qquad 12.1.15$$

For a mixture of galena and quartz particles settling in a pulp of density 1.5, the hindered settling ratio is:

$$\frac{7.5 - 1.5}{2.65 - 1.5} = 5.22$$

i.e., a galena particle will settle at the same rate as quartz particle of diameter 5.22 times larger than galena particle.

When the hindered settling ratio of 5.22 is compared with the free settling ratio of 3.94, it is evident that hindered settling reduces the effect of size, while increasing the effect of density, which means that heavier (or lighter) particles can be separated in hindered settling.

Hindered settling ratio is always greater than the free settling ratio. As the pulp density increases, this ratio also increases.

Free settling conditions are used in classifiers, in which case they are called Free settling classifiers (Mechanical classifiers or Horizontal current classifiers), to increase the effect of size on separation. Hindered settling conditions are used in classifiers, in which case they are called Hindered settling classifiers (Hydraulic classifiers or Vertical current classifiers), to increase the effect of density on separation.

The motion of a particle, when it starts settling in the fluid and move through the fluid, can be divided into two stages: the acceleration period and the terminal velocity period. Initially the velocity of the particle is zero with respect to the fluid and increases to the terminal velocity during a short period, usually of the order of one tenth of a second or less. During this first stage of short period, there are initial-acceleration effects. Once the particle reaches its terminal velocity the second stage starts and continues as long as the particle continues to settle. Classification and thickening processes make use of the terminal velocity period.

In jigging operation, particles are allowed to settle during the acceleration period.

Chapter 13

Classification

Classification is a method of separating mixtures of particles of different sizes, shapes and specific gravities into two or more products on the basis of the velocity with which the particles fall through a fluid medium i.e., settling velocity. Generally classification is employed for those particles which are considered too fine to be separated efficiently by screening.

The basic principle of classification is:

The coarser, heavier and rounder particles settle faster than the finer, lighter and more angular particles

In classification, certain particles are only allowed to settle in the fluid medium in order to separate the particles into two fractions.

13.1 CLASSIFIERS

The units in which the separation of solids in fluid medium is carried out are known as classifiers. These classifiers may be grouped into three broad classes as:

1 Sizing classifiers.
2 Sorting classifiers.
3 Centrifugal classifiers.

Let us reconsider Figure 12.1 by assigning the numbers in the order of increasing size as shown in Figure 13.1 to explain how the particles can be separated.

Under free settling conditions, if sufficient time for the light particle 5 is not given for settling, all the light particles of 1 to 5 and heavy particles of 1 to 4 will be at the top of the classifier, and light particles of 6 to 8 and heavy particles of 5 to 8 will be at the bottom of the classifier. Top and bottom fractions are removed by suitable means without giving time for the light particle 5 to settle. Thus the two fractions obtained from the classifier contain:

Top fraction – light particles of 1–5 and heavy particles of 1–4.
Bottom fraction – light particles of 6–8 and heavy particles of 5–8.

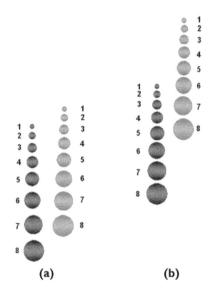

Figure 13.1 (a) Free settling; (b) Hindered settling.

Each fraction contains both light and heavy particles and almost the same size or closely sized particles. It means that all the particles are separated into two size fractions. Hence this type of classification is called sizing classification.

In case of hindered settling conditions, if top and bottom fractions are removed without allowing light particle 8 to settle, the bottom fraction contains solely of heavy particles 6 to 8. As the settling velocity of all the light particles is less than the settling velocity of light particle 8, they remain at the top along with the heavy particles 1 to 5 and get discharged as overflow product. If these particles are again classified without allowing light particle 5 to settle, the bottom fraction contains heavy particles 1 to 5 and light particles 6 to 8. Only light particles 1 to 5 remain at the top and get discharged as overflow product.

It is obvious that heavy particles of 1 to 5 and light particles of 6 to 8 are under hindered settling conditions, which means that they cannot be separated. Heavy particles of 6 to 8 can be separated as bottom fraction and light particles of 1 to 5 can be separated as top fraction.

It is very interesting to note that, under free settling conditions, only heavy particle 8 and light particles 1 and 2 can be separated whereas in hindered settling conditions, heavy particles of 6 to 8 and light particles of 1 to 5 can be separated.

From this example, it can be reiterated that free settling conditions are to be maintained to separate the mixed density and mixed size particles according to their size, which increases the effect of size and decreases the effect of density on separation. Hindered settling conditions are to be maintained to separate the mixed density and mixed size particles according to their densities which increases the effect of density and decreases the effect of size on separation.

13.1.1   Sizing classifiers

A typical sizing classifier consists of a sloping rectangular trough. Feed slurry is introduced at point 1 as shown in Figure 13.2 fines overflow at point 2. The rate of feed and distance between point 1 and point 2 is selected in such a way that the rate of travel of required fine particles must be more than the rate of their settling so that all the required fine particles overflow at point 2. The coarse particles settle to the bottom. To remove these coarse particles, a mechanical means such as spiral or rake is placed at the bottom of the trough. Figure 13.3 shows how the particles are separated in Spiral and Rake classifiers.

Sizing classifiers are the mechanical classifiers. Since the stream of slurry consists of fines flow horizontally from the feed inlet to the overflow weir, these are also called as horizontal current classifiers. They are also called pool classifiers as the classification takes place in the pool. These classifiers are extensively used in closed circuit grinding operation with a ball mill where underflow coarse product is directly fed to the inlet of the ball mill. Another type of classifier is of tank type such as Dorr Bowl.

These classifiers are:

- Free settling classifiers.
- Uses relatively dilute aqueous suspension.
- Perform mostly sizing.
- Percent solids are usually 5–10%.
- Yields only two products.

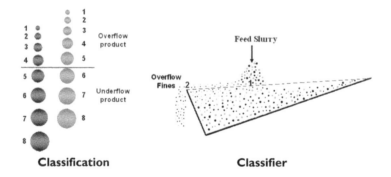

Classification   **Classifier**

Figure 13.2 Principle of a mechanical classifier.

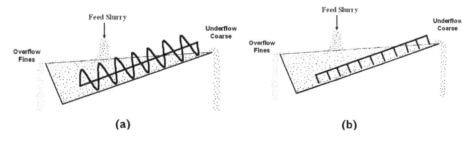

(a)   **(b)**

Figure 13.3 (a) Spiral classifier; (b) Rake classifier.

13.1.2 Sorting classifiers

Sorting classifiers employ the hindered settling conditions to increase the effect of density in order to separate the particles according to their density rather than size. A typical sorting classifier consists of a series of sorting columns as shown in Figure 13.4.

The feed slurry is introduced centrally near the top of the first sorting column. A current of water known as **hydraulic water** is introduced at the bottom of the sorting column at a velocity slightly less than the smallest heavy particle among the particles required to be discharged in the first sorting column. All those particles having settling velocity less than that of rising water velocity will not settle and rise to the top of the column and fed to the second column. The particles, having a settling velocity more than that of the rising water velocity, settle to the bottom of the first sorting column and get discharged through the spigot. The velocity of the hydraulic water in the second sorting column is less than that of the velocity in the first sorting column so that the particles of low settling velocity settle to the bottom of the second sorting column and get discharged through the spigot. Similarly the particles with still low settling velocity are obtained through the spigot of third sorting column and remaining particles are obtained as overflow from the third sorting column.

Explanation with reference to Figure 13.1 has already been given as to how the particles are separated under hindered settling conditions by using two sorting columns where it is clear that fine heavy and coarse light particles are together discharged as spigot product of the second sorting column, coarse heavy as spigot product of the first sorting column and fine light as overflow product of the second sorting column. Figure 13.5 shows this separation. If the free settling conditions are maintained, a series of spigot products with decreasing size of particles from first spigot are obtained as shown in Figure 13.6.

As these classifiers use the rising current of water, they are called hydraulic classifiers and vertical current classifiers.

These classifiers are:

• launder type with rectangular boxes attached to it such as the Evans classifier, cylindrical type such as the Anaconda and Richards classifiers, trapezoidal tank type such as the Fahrenwald sizer.

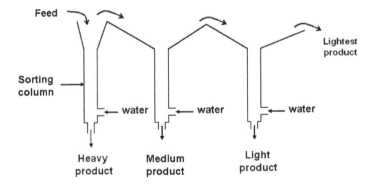

Figure 13.4 Principle of sorting classifier.

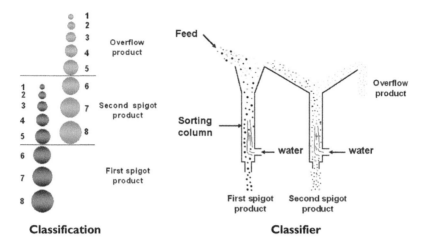

Figure 13.5 Hydraulic classifier with sorting effect.

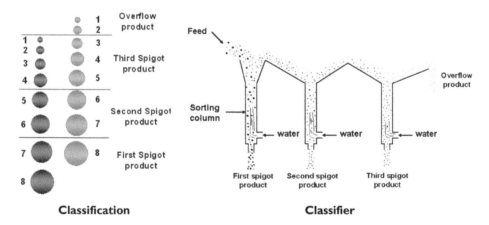

Figure 13.6 Hydraulic classifier with sizing effect.

- Uses rising current of water called hydraulic water.
- Hindered settling classifiers.
- Uses relatively dense aqueous suspension as fluid medium.
- Perform mostly sorting.
- Percent solids are usually 15–30%.
- Yield more products.

Even though sorting classifiers are not truly sizing classifiers, they are sometimes used to sort out the particles in a close size range as shown in Figure 13.6 which are necessary for gravity concentration operations such as tabling. The Stokes Hydrosizer is commonly used to sort the feed to gravity concentrators.

13.1.3 Centrifugal classifiers

Under gravity force, the settling rate of a particle varies as its effective mass. If a centrifugal force is applied, the effective mass increases and therefore settling rate increases. As particles are ground smaller they reach a size where the surface drag against the surrounding fluid almost neutralizes the gravitational pull, with the result that the particle may need hours, or even days, to fall a few inches through still water. This slowing down of the settling rate reduces the tonnage that can be handled and increases the quantity of machinery and plant required. By superimposing centrifugal force, the gravitational pull can be increased from 50 to 500 times depending on the pressure at which the pulp is fed and the size of the vessel. The hydrocyclone is one which utilizes centrifugal force to accelerate the settling rate of particles.

Hydrocyclone (Figure 13.7) has no moving parts. It consists of a cylindrical section with a tangential feed inlet. A conical section, connected to it, is open at the bottom, variously called the underflow nozzle, discharge orifice, apex or spigot. The top of the cylindrical section is closed with a plate through which passes an axially mounted central overflow pipe. The pipe is extended into the body of the cyclone by a short, removable section known as vortex finder, which prevents short-circuiting of feed directly into the overflow.

When a pulp is fed tangentially into a cyclone, a vortex is generated about the longitudinal axis. The accompanying centrifugal acceleration increases the settling rates of the particles, the coarser of which reach the cone's wall. Here they enter a zone of reduced pressure and flow downward to the apex, through which they are discharged.

At the center of the cyclone is a zone of low pressure and low centrifugal force which surrounds an air-filled vortex. Part of the pulp, carrying the finer particles with major portion of feed water, moves inward toward this vortex and reaches the gathering zone surrounding the air pocket. Here it is picked up by the vortex finder, and removed through a central overflow orifice (Figure 13.8).

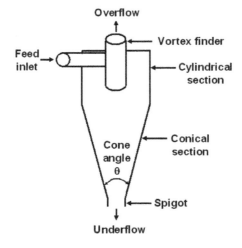

Figure 13.7 Hydrocyclone.

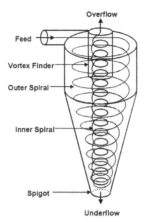

Figure 13.8 Hydrocyclone operation.

The size of the hydrocyclone is the diameter of its cylindrical section. The variables that affect the performance of a hydrocyclone can be divided into two groups as design variables and operating variables. Design variables are the size of the hydrocyclone, diameter of feed inlet, vortex finder and apex, and position of the vortex finder. Operating variables are feed rate, feed pressure, solid-liquid ratio, density, size and shape of feed solids, and density and viscosity of liquid medium.

Advantages of the Hydrocyclone are:

1 Sharper classification.
2 Saving of floor space.
3 Less power consumption.
4 Less maintenance.
5 Ability to shut down the mill immediately under full load.
6 Ability to bring the circuit rapidly into balance.
7 Elimination of cyclic surging.

The main use of the hydrocyclone in mineral beneficiation is as a classifier, which has proven extremely efficient at fine separation sizes (between 150 and 5 microns). It is used increasingly in closed-circuit grinding operations. It is also used for many other purposes such as de-sliming, de-gritting and thickening. It has also found wide acceptance for the washing of fine coal in the form of Heavy Medium Cyclone and Water Only Cyclone.

Chapter 14

Beneficiation operations

Beneficiation operations are physical or mechanical unit operations where valuable mineral particles are separated from the mixture of liberated valuable mineral particles and gangue mineral particles. Every beneficiation operation is based on one or more physical properties in which valuable and gangue minerals differ in that property.

Major beneficiation operations and the basis of separation are given in Table 14.1 The following important terminology is used in Mineral Industries.

Concentrate: The valuable mineral product obtained from the beneficiation operation.
Tailing: The waste or gangue mineral product obtained from the beneficiation operation.

Practically all the valuable mineral particles cannot be separated due to:

* Incomplete liberation (degree of liberation <100%).
* Inefficiency of beneficiation operation (efficiency of operation <100%).

Hence both the concentrate and the tailing contain locked particles of valuable and gangue minerals and the concentrate contains a little amount of gangue mineral particles and the tailing contains a little amount of valuable mineral particles.

Middling: If the valuable mineral particles are not completely liberated during the size reduction operations, (i.e. degree of liberation is less than 100%) some particles contain both valuable and gangue minerals. Such particles can be separated in a

Table 14.1 Beneficiation operations and basis of separation.

Beneficiation operations	Basis of separation
Hand Sorting	Colour & Lustre
Mechanical Sorting	Friction & Shape
Gravity Separation	Sp.gr, Size & Shape
Centrifugal Separation	Sp.gr, Size & Shape
Flotation	Wettability & Sp.gr
Magnetic Separation	Magnetic susceptibility
Electrical Separation	Electrical conductivity

beneficiation operation as a third product. This third product is called middling. Usually, this middling product is comminuted for further liberation and then beneficiated.

Simple expressions are being used for the performance evaluation of beneficiation operations and controlling those operations in mineral industries. For a two product beneficiation operation:
Let:

F = Weight of the feed
C = Weight of the concentrate
T = Weight of the tailing
f = Fraction of the metal (or material) present in the feed
c = Fraction of the metal (or material) present in the concentrate
t = Fraction of the metal (or material) present in the tailing

then the material balance for total material is:

$$F = C + T \qquad\qquad 14.1$$

i.e., total material input = total material output
Similarly, the material balance for valuable metal (or material) they contain is:

$$Ff = Cc + Tt \qquad\qquad 14.2$$

i.e., valuable metal (material) input = valuable metal (material) output

Ratio of concentration: It is defined as the ratio of the weight of the ore fed to the weight of the concentrate produced.

$$\text{Ratio of concentration} = \frac{F}{C} \qquad\qquad 14.3$$

From the material balance equations 14.1 and 14.2, ratio of concentration can be computed in terms of assay values as:

$$\text{Ratio of concentration} = \frac{c-t}{f-t} \qquad\qquad 14.4$$

Ratio of recovery: Defined as the fraction (or percent) of the total amount of valuable metal (material) originally present in the ore which is recovered in the concentrate.

$$\text{Ratio of recovery} = \frac{Cc}{Ff} \qquad\qquad 14.5$$

From the material balance equations 14.1 and 14.2, ratio of recovery can be computed in terms of assay values as:

$$\text{Ratio of recovery} = \frac{c(f-t)}{f(c-t)} \qquad 14.6$$

Ratio of enrichment: It is defined as the ratio of the assay value of the concentrate to that of the ore.

$$\text{Ratio of enrichment} = \frac{c}{f} \qquad 14.7$$

This is equal to the product of the ratio of concentration and the recovery. The ratio of enrichment does not have the direct economic significance of the commonly used ratio of concentration. It gives better idea of the cleanliness of the concentrate.

Metallurgical efficiency: It is defined as the arithmetical average of the recoveries of the principal constituent of each product (including tailing).

$$\text{Let } R_v = \text{recovery of valuable in concentrate} = \frac{Cc}{Ff}$$

$$J_w = \text{rejection of waste in tailing} = \frac{T(1-t)}{F(1-f)}$$

$$\text{Then, Metallurgical efficiency} = \frac{R_v + J_w}{2} \qquad 14.8$$

Economic recovery or efficiency: It is the ratio of the actual value of the concentrate obtained from a ton of ore to the value of the concentrate theoretically obtainable in mineralogically pure form from a ton of ore.

If 0.20 ton of lead concentrate worth Rs.10,000/- per ton is obtained in practice from an ore that theoretically should yield 0.18 ton of mineralogically pure galena concentrate worth Rs.15,000/- per ton,

$$\text{Economic recovery} = \frac{0.20 \times 10000}{0.18 \times 15000} = 0.741 = 74.1\%$$

To achieve the recovery to the maximum possible extent, the ore is repeatedly treated by using another unit of the same device. The following is the terminology in use in such cases:

Roughing: It is the operation of removal of a rough concentrate at the earliest stage of treatment of the ground ore. The device used is called **Rougher.**

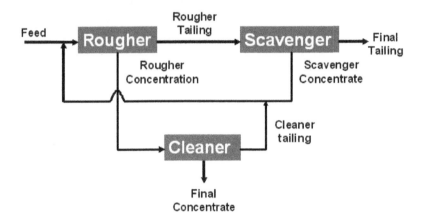

Figure 14.1 Flow sheet of three stage treatment.

Scavenging: It is the operation of removal of the last recoverable fraction of valuables before discarding the final tailing from the treatment plant. The device used for this purpose is called **Scavenger.**

Cleaning: It is the operation of re-treating the rough concentrate to improve its quality. The device used for this cleaning stage is called **Cleaner.**

Figure 14.1 is a typical flow sheet showing a three stage treatment.

Chapter 15

Gravity concentration

Gravity Concentration is a method of separating the mineral particles based on their specific gravities when they are allowed to settle in a fluid medium.

Some idea of the type of separation possible can be gained from the ratio, called **concentration criterion**, and is defined as:

$$\frac{\rho_b - \rho_f}{\rho_l - \rho_f}$$

where ρ_b is the density of the heavy mineral, ρ_l is the density of the light mineral, and ρ_f is the density of the fluid medium. This ratio is a comparison of the buoyancy forces which are at work.

In general, when this ratio is greater than 2.5, whether positive or negative, then gravity separation of particles down to 75 micron size is possible. As the value of this ratio decreases, the efficiency of separation decreases, and if it is below 1.25, gravity concentration is not commercially feasible in general. The following are the values of ratios and particle sizes based on experimental data:

Concentration criterion Suitability to gravity separation

>2.50	easy down to 75 μm (200 mesh)
1.75–2.50	possible down to 150 μm (100 mesh)
1.50–1.75	possible down to 1.7 mm (10 mesh)
1.25–1.50	possible down to 6.35 mm
<1.25	no separation is possible in water at any size
	Another (heavier) fluid, or a pseudo fluid medium heavier than H_2O, is needed. An alternative is to alter the effective particle density by causing it to adhere to an air bubble as in flotation.

Some concentration criterion ratios for minerals that are treated by gravity separation are given in Table 15.1.

The motion of a particle in a fluid depends not only on its specific gravity but also on its size; large particles will be affected more than smaller ones. The efficiency of gravity concentration operation, therefore, increases with particle size. Smaller particles respond poorly because their movement is mainly dominated by surface

Table 15.1 Concentration criterion of minerals separated by gravity separation from a gangue of density 2.65 gm/cc.

Mineral	Fluid	Concentration criterion
Gold	Water	10.3
Gold	Air	6.8
Cassiterite	Water	3.5
Coal	Water	3.4
Hematite	Water	2.5

friction. In practice, close size control of feeds to gravity processes is required in order to reduce the size effect and make the particles move depending on their specific gravities.

15.1   FLOAT AND SINK

Float and sink, also called Heavy Liquid separation (HLS), is an operation where particles of different specific gravities are separated by using a suitable heavy liquid. The principle of float and sink is:

> When two particles of different specific gravities are introduced in a liquid having specific gravity intermediate between that of the two particles, the lighter particle would float and the heavier particle would sink.

In principle, it is the simplest and standard laboratory method for separating minerals of different specific gravities. Solutions of inorganic salts like calcium chloride and zinc chloride are used as heavy liquid. Table 15.2 shows some of the organic liquids that are used for separation of mineral particles.

Solutions of required specific gravity are prepared by adding two or more liquids at different proportions. When the ore particles are introduced into this solution, mineral particles of less specific gravity will float and mineral particles of a more specific gravity will sink. Then two products are removed from the solution.

Separation of raw coal into different specific gravity fractions and separation of minerals, Ilmenite, Monazite, Rutile, Zircon, Garnet from Beach Sands, are the two important examples for the application of float and sink.

The buoyant forces acting on the light particles in a dense medium make them rise to the surface and the dense particles sink to the bottom because of gravity force. In a static bath, force balance equation is written as:

$$F_g = (m_p - m_f)\, g \qquad\qquad 15.1.1$$

where

F_g = gravitational force
m_p = mass of the solid particle
m_f = mass of the fluid displaced by the particle

Table 15.2 Organic liquids and their specific gravities.

Liquid	Chemical formula	Specific gravity
Benzene	C_6H_6	0.80
Carbon tetrachloride	$C\,Cl_4$	1.58
Pentachloro ethane	$C\,Cl_2CH\,Cl_2$	1.67
Methylene Bromide	CH_2Br_2	2.48
Bromoform	$CH\,Br_3$	2.89
Tetrabromo ethane	$C_2H_2Br_4$	2.96
Methylene iodide	CH_2I_2	3.31
Thallous formate solution	$H\,COOTI$	3.39

For the particles which float, F_g will have a negative value and for the particles which sink, F_g value is positive. In a centrifugal separator, equation 15.1.1 becomes:

$$F_C = (m_p - m_f)\frac{v^2}{R}$$

15.1.2

where
F_C = centrifugal force
v = tangential velocity
R = radius of the centrifugal separator

From these two equations, it is clear that the forces causing the separation of the particles in a static bath are proportional to g whereas in a centrifugal separator, separating forces are proportional to $\frac{v^2}{R}$ which is much greater. Hence particles down to 0.5 mm size can be separated by centrifugal separators.

Heavy Liquid Separation is not carried out now-a-days in mineral or coal industries because of the following reasons:

1   Some liquids are toxic, some others are corrosive.
2   Organic liquids are expensive.
3   They are absorbed by particle surfaces and contaminate the mineral particles.
4   Loss of liquid is high, due to volatility.
5   These liquids cannot be recovered economically.
6   It is not always possible to get hold of the required heavy liquids.

An alternative process known as **Heavy Medium Separation (HMS)** or **Dense Medium Separation (DMS)** is adopted in industries.

15.2   HEAVY MEDIUM SEPARATION

Heavy Medium Separation or Dense Medium Separation is a process similar to Heavy Liquid Separation but instead of heavy liquid, a pseudo liquid is used. A **pseudo liquid** is a suspension of water and solids which behaves like a true liquid. The solids used are known as **medium solids** and they must meet the following requirements:

1   They must be sufficiently high in density.
2   They must be physically strong.
3   They must be chemically inert.
4   They must be readily available at a low cost.
5   They must be easily recoverable.
6   They must not enter into the cracks of cleaned ore lumps.
7   The resulting viscosity should be low.

In order to produce a stable suspension of sufficiently high density, with a reasonably low viscosity, it is necessary to use fine, high specific gravity solid particles. Agitation is necessary to maintain the suspension and to lower the apparent viscosity.

Some of the medium solids employed in industries for making suspensions are given in Table 15.3.

The percent medium solids (C_w) to be added to maintain the suspension at required specific gravity can be calculated from the equation 11.3:

$$C_w = \frac{100\rho_p (\rho_{sl} - 1)}{\rho_{sl}(\rho_p - 1)}$$

Heavy Medium Separation is applicable to any ore. It is widely used for washing coal at coarser sizes. Chance Cone, used for washing coal of 75–13 mm size is the oldest HMS unit. Barvoys Process, Dutch State Mines Process, Tromp Process, Drewboy Process, and Wemco Drum Separation Process are the other Heavy Medium Separation Processes employed in Coal Washeries.

15.2.1   Chance cone process

The vessel of a Chance cone process, consists of an inverted cone with a cylindrical upper section, has two rotating paddles at its center to maintain the suspension and

Table 15.3 Medium solids.

Solids	Specific gravity
Sand	
Loess	2.6
Shale	
Barytes	4.0
Magnetite	5.0
Ferrosilicon	7.0
Galena	7.5
Some clays	–

a refuse chamber to discharge refuse material without disturbing the contents of the vessel. Raw coal is fed into the top of the vessel at one side and the clean coal that floats at the surface is discharged from the other side. Sand, used as medium solids, is added to the vessel continuously in the form of a thick pulp to keep the level of the suspension up to the level of the overflow lip at the top of the vessel. The water required for the agitation and maintenance of the suspension is introduced through water inlets at various levels in the cone. The clean coal and refuse discharged from the chance cone are de-sanded and de-watered. The sand and water thus recovered are reused through recirculation.

When it is desired to obtain the middling product in addition to the clean coal and refuse products, a separate middlings column is provided beside the cone. The chance cone can be operated as a single gravity separator or a two gravity separator merely by opening or closing the valves controlling the sand and water for lifting the middlings up the column. Figure 15.1 shows the three product chance cone.

15.2.2 Centrifugal separators

Heavy medium cyclone is a centrifugal separator similar to the conventional hydrocyclone in principle of operation. It provides a high centrifugal force and a low viscosity of the medium. Much finer separations are achieved in Heavy medium cyclone than in gravity separators. The ore or coal is suspended in a very fine medium of ferrosilicon or magnetite, and is fed tangentially to the Heavy medium cyclone under pressure. The sink product leaves the cyclone at the apex and the float product is discharged through the vortex finder. These cyclones are commonly installed with axes at 10–15° to the horizontal, thereby enabling the unit to be fed at comparatively low inlet pressure, preferably from a steady head tank. Heavy medium cyclones are widely used for beneficiation of coal in the size range of 13–0.5 mm in coal washeries.

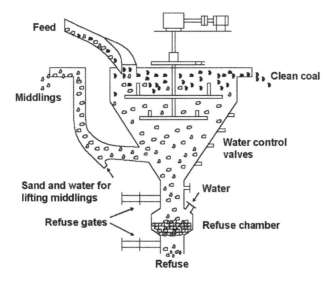

Figure 15.1 Chance cone separator.

By giving a special shape to the cone, and to the cylindrical part of the cyclone, it is possible to affect separation without using heavy medium. Such a cyclone is known as **water-only cyclone** or **water washing cyclone**. Vorsyl separator, LARCODEMS, Dyna Whirlpool and Tri-Flo separator are some other centrifugal separators.

15.3 JIGGING

Jigging is the process of separating the particles of different specific gravities, size and shape by introducing them on a perforated surface (or screen) through which the water is made to flow by pulsion and suction strokes alternatively.

In jigging, the particles are allowed to settle for only a short period and the particles will never attain their terminal velocities. It means that separation will depend on the initial settling velocities of the particles. The particles will settle during their accelerating period. The initial settling velocity is extremely low and the drag force due to frictional effects is not developed. Under these circumstances, the drag force is practically zero and the two principal forces acting on the particle are the gravitational and the buoyant forces. Then the force balance equation becomes:

$$m_p \frac{dv}{dt} = (m_p - m_f)g \qquad\qquad 15.3.1$$

$$\frac{dv}{dt} = \frac{m_p - m_f}{m_p} g = \left[1 - \frac{\rho_f}{\rho_p} \right] g \qquad\qquad 15.3.2$$

Equation 15.3.2 shows that the initial acceleration of the particles during settling depends on the force of gravity, density of the particle and the fluid and does not depend on the size or shape of the particle. Initial acceleration is the maximum for the densest particles. This situation indicates that light and heavy particles can be separated by providing short duration of settling. It can also be explained by considering three heavy particles and three light particles of different sizes. The six curves in Figure 15.2 shows the settling velocities of these particles as a function of time.

If the particles are allowed to settle for time t_3 small heavy and coarse light particles settle equally. Similarly fine heavy and small light particles settle equally but with different velocities. Coarse heavy particles settle faster than all other particles and fine light particles settle slower than all other particles. Hence only these two particles can be separated.

If the time of settling is reduced to t_2, it is not possible to separate fine heavy and coarse light particles as they fall to an equal distance in a settling period of t_2. However coarse and small heavy particles can be separated as they settle faster than all other particles. Similarly small and fine light particles can be separated as they settle slower than all other particles.

If the time of settling is still reduced to t_1, all heavy particles will fall to a greater distance than all light particles and hence all heavy and light particles can be completely separated irrespective of their size.

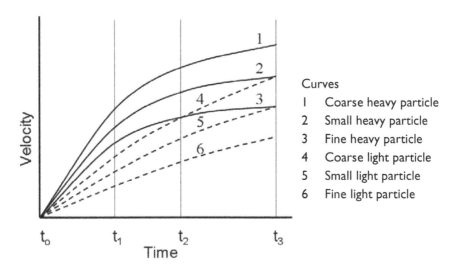

Figure 15.2 Settling velocities of six particles.

The short settling time t_1 is achieved in jigging by giving frequent pulsion and suction strokes of a fluid to a bed of mixed particles of different densities and sizes. After repeated pulsion and suction strokes of the fluid, the particles are rearranged in a bed according to their densities and sizes.

Let a simple experiment with a circular screen fixed inside a vertical hallow cylinder, which can be called a Jig, be considered. A group of light particles, all of the same size, with one heavy particle at the top of light particles are placed on a screen as shown in Figure 15.3(A). If water is introduced from the bottom to create a pulsion stroke, all the particles are pushed upwards. Light particles are pushed more along the distance while heavy particles are pushed less along the distance as shown in Figure 15.3(B). Now if the water is withdrawn from the bottom to create a suction stroke, light particles settle down less along the distance while the heavy particles settle down more along the distance. This results in the rearrangement of the particles as shown in Figure 15.3(C) where heavy particles are placed one layer below from the top. During the further repeated pulsion and suction strokes, the same process is repeated as shown in Figure 15.3(D) to (G) and at the end, the heavy particles are placed in the bottom layer. The particles settle by **hindered settling** as a smaller quantity of water is present so that heavy particles are always placed at the bottom.

If a group of both light and heavy particles, all of same size, are taken in equal number as shown in Figure 15.4(A) and do the same experiment, all the particles are rearranged as shown in the Fig. 15.4(B) after several pulsion and suction strokes. Rearrangement of particles is called **stratification**.

If a group of both light and heavy particles of different sizes are used for the same experiment, all the particles are rearranged in similar fashion. But small particles pass through the interstices between coarse particles as shown in Figure 15.5. This phenomenon is called **consolidation trickling**.

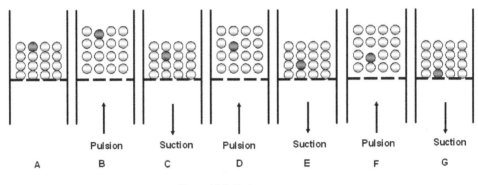

Figure 15.3 Jigging process.

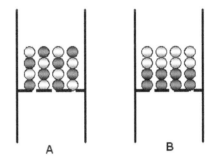

Figure 15.4 Stratification of equal size particles due to jigging action.

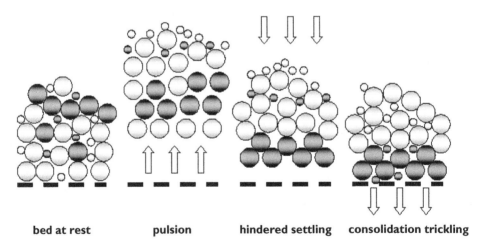

Figure 15.5 Stratification of different size particles due to jigging action.

If repetition of fall is frequent enough, and the duration of fall is short enough, the difference in distance traveled by dissimilar particles will be more due to the difference in their initial acceleration. Under these conditions, stratification would take place on the basis of density alone. This is the reason why the particles, having little density difference, are stratified effectively in jigging.

The following three effects contribute to the stratification in jigs:

1   Hindered settling classification.
2   Differential acceleration at the beginning of fall.
3   Consolidation trickling at the end of fall.

During the pulsion stroke, the solid bed is opened and expanded. When pulsion ceases, the particles settle into more homogeneous layers under the influence of gravity during the suction stroke. Stratification during the stage that the bed is open is essentially controlled by hindered settling classification, as modified by differential acceleration, and during the stage that the bed is tight it is controlled by **consolidation trickling.** The first process puts the coarse-heavy particles on the bottom, the fine-light particles at the top, and the coarse-light and fine-heavy particles in the middle. The second process puts the fine-heavy particles at the bottom, the coarse-light particles at the top, and the coarse-heavy and fine-light particles in the middle. By varying the relative importance of the two, and by varying the importance of differential acceleration, an almost perfect stratification according to density alone can be obtained. The frequency of pulsations usually varies from 50–300 cycles per minute. Settling ratio in jigging, called **jigging ratio,** is more because the suspension is so thick.

Hydraulic jigs use water as the fluid medium whereas pneumatic jigs use air. The device used for the jigging process is called **Jig.** The basic construction of a Hydraulic jig is shown in Figure 15.6.

It consists of a rectangular open top container with the screen at the bottom which supports the bed of solids. Pulsion and suction strokes of water through the screen are given by means of a reciprocating piston or compressed air. The feed is introduced at one end of the jig and products are taken out by two methods. In one method, both concentrate and tailing are removed from the jig bed. This method is called as **Jigging on the screen.** In another method, called **Jigging through the screen,**

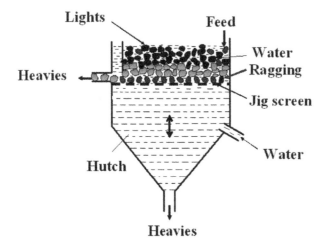

Figure 15.6 Basic construction of a hydraulic jig.

light product is removed from the jig bed whereas heavy product is removed from below the jig screen i.e., from the bottom of the jig box known as **hutch**. In such a case, heavy product is known as **hutch product** which passes through jig screen. A mineral bed, called **ragging** material, is sometimes placed on the jig screen to permit the heavy product to pass through the screen and to prevent the passage of light product into the hutch.

Jigs are of two types namely **Fixed screen jigs** and **Movable screen jigs**. In Fixed screen jigs, water is made to pulsate whilst the screen is stationary whereas in Movable screen jigs, the screen moves downward and upward in a stationary fluid and produces pulsion and suction strokes. The Harz jig and Baum jig are the two important Fixed screen jigs in use.

In Harz jig (Figure 15.7), the piston moves up and down creating the necessary pulsations of water in the jig compartment. A number of compartments are placed in series for successive stages of separation of products of different qualities. The feed is introduced at one end of the first compartment. The light fraction stratifies upward and moves into the second compartment for further separation. The final light fraction overflows from the last compartment. The heavy fractions are removed continuously through a discharge gate at the side of each compartment. Even though the Harz jig is essentially on the screen jig, some fine concentrate will inevitably find its way to the hutch, where it is periodically removed by manual methods. Harz jigs are used in the processing of metallic ores in the size range of 37 mm to 0.5 mm.

The Baum jig (Figure 15.8) is the basis of most of the jigs commonly used in coal washing industry and consists of a U-shaped steel container divided into two in cross section and longitudinally divided into two or more compartments. On one side of the jig is a perforated screen and other side is a pulsion chamber fitted with an air valve, connected to a compressed air supply which produces water pulsations at the rate of 30 to 60 cycles/min. The lights are discharged over the weir and the heavies are removed through a discharge gate and/or drop into the hutch compartment and raised by a bucket elevator out of the jig. The Baum jig is used for any size range of a coal below 20 cm.

The Denver Mineral Jig, Circular Jig (also known as IHC Cleaveland Jig), Kelsey Centrifugal Jig, Inline Pressure Jig, Batac jig, and Feldspar Jig are some of the other Jigs.

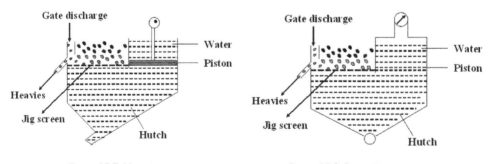

Figure 15.7 Harz jig.   Figure 15.8 Baum jig.

15.4 FLOWING FILM CONCENTRATION

Flowing film concentration has been defined as sorting of mineral particles on flat surfaces in accordance with the size, shape and specific gravity of the particles moved by a flowing film of water.

When water is made to flow over a bare sloping deck, the velocity of water adjacent to the deck is zero and increases as the distance from the deck increases reaching maximum at the top surface of water (Figure 15.9). However, velocity at the top surface of water is slightly less than the maximum due to air friction.

If a number of spheres, composed of two kinds of minerals, one heavy another light, and are of different sizes, are introduced into a thick layer of water, they will be separated during their fall through this layer. The biggest heavy sphere falls faster on to the deck through water and is least affected by the current and lies nearest to the point of entry. The smallest light sphere will drift furthest downstream. The others will drift to different distances. The particles' drift in flowing water is shown in Figure 15.10.

The flowing water presses the sphere and makes to move downstream. The differential rate at which the water is flowing over the deck causes low pressure on the

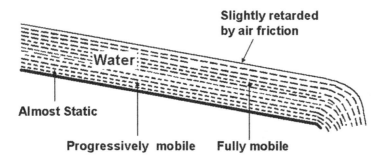

Figure 15.9 Flow of water on sloping deck.

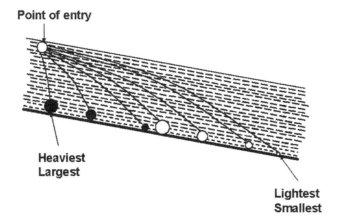

Figure 15.10 Particles' drift in flowing water.

bottom of the sphere tending to slide on the deck and causes high pressure at the top of the sphere tending to roll on the deck. Figure 15.11 shows the various forces applied by flowing water on a submerged particle.

Since small particles are submerged in the slower-moving portion of the film, they will not move as rapidly as coarse particles. If the combined influence of deck slope and streaming velocity is sufficient to keep all the spheres in rolling movement, they rearrange themselves in the following downslope sequence (Figure 15.12)

1   Small-heavy particles.
2   Coarse-heavy and small-light particles.
3   Coarse-light particles.

It is to be noted that in flowing film concentration coarse-heavy particles are placed with small-light particles which is the reverse of the stratification that takes place in classification.

Pinched sluice (Figure 15.13) is an inclined stationary trough 60 to 90 cm long, narrowing from about 24 cm in width at the feed end to 3 cm at the discharge end. Feed consisting of 50–65% solids enters the sluice and stratifies as the particles flow through the sluice and crowds into the narrow discharge area. The crowding causes the bed to dilate allowing heavy minerals to migrate and move along the bottom, while lighter particles are forced to the top. The resulting mineral bands are separated by a splitter at the discharge end. Pinched sluices are simple devices and are inexpen-

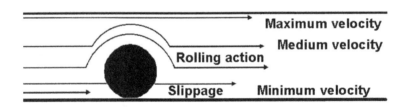

Figure 15.11 Forces of flowing water on particle.

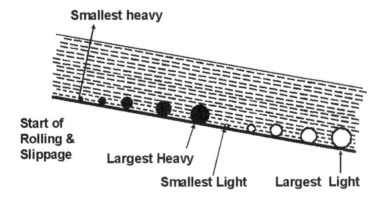

Figure 15.12 Arrangement of particles over a deck of flowing film.

sive to build and run. Pinched sluices are mainly used for separation of heavy mineral sands.

Reichert Cone (Figure 15.14) is based on the pinched sluice concept but employs an inverted cone instead of a rectangular channel. The crowding and dilating effect of the bed is produced by a reduction in perimeter as the material approaches the center discharge point. Reichert cones are more efficient than pinched sluices because there are no sidewalls to interfere in the separation process.

Inclined stationary tables are another means of stationary flowing film concentration and are also known as buddles. These consist of an inclined table in which crushed ore is washed with running water to flush away impurities. The pulp is fed at the upper end which flows down over the table surface. The slope is so adjusted that the lighter particles get washed off, while the heavier ones settle and accumulate on the surface. Carduroy table is popular for recovery of gold.

Humphrey Spiral (Figure 15.15) is another typical unit of stationary flowing film concentrator. It consists of a descending spiral launder of a helical conduit with a modified semi-circular cross section. When the slurry is made to flow over a spiral launder, centrifugal force comes into play in addition to gravity and frictional drag

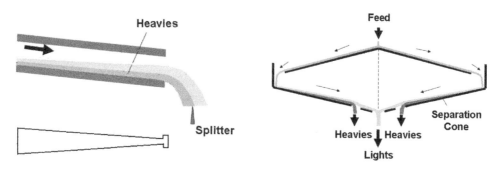

Figure 15.13 Pinched sluice.   Figure 15.14 Reichert cone.

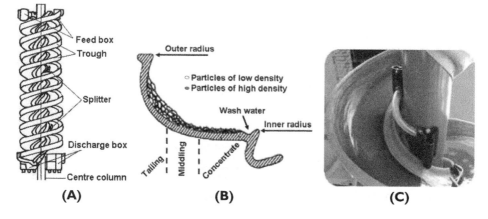

Figure 15.15 Humphrey spiral concentrator. (Courtesy Outotec (USA) Inc.)

forces. As a result, the heavies flow towards the centre and the lights are thrown away from the centre and flow towards the outer edge of the launder. Wash water runs down a small trough alongside the inner horizontal part of the cross section and can be deflected into the portion carrying the pulp whenever required. At intervals, withdrawal ports are provided at the lowest points in the cross section down the spiral. Adjustable splitters are arranged at the ports to control the removal of part of the inner pulp stream which is further discharged to the central collecting pipe. Figure 15.15 (B) shows how stratification takes place. Figure 15.15 (C) shows close-up view of the spiral with wash water addition and heavy particles splitting. Humphrey spiral is extensively used in the beach sand industry to recover the heavies and discard the lights. They are effective for particles in the range of 3 mm to 75 μm and for minerals with specific gravity differences greater than 1.0.

If an obstruction, like cross-riffles (bars set across the stream), is placed on a bare sloping deck, this obstruction introduces turbulent flow, and vortices; the smallest and heaviest particles occupy the bottom place and are covered by heavy particles of increasing size and then small lighter particles. At the top of the bed are coarsest light particles which may pass over the obstruction (Figure 15.16).

If a reciprocating motion is given to the deck, the particles in between the successive riffles are subject to hindered settling and consolidation trickling and move to one side of the deck whereas lighter particles are washed off across the riffles and along the streaming current. This operation is called as **tabling**. Figure 15.17 shows how the particles are stratified in between the riffles as a result of the shaking motion in a tabling operation.

Fig. 15.17 clearly indicates that coarse lights and fine heavies can only be separated as individual fractions (products) and fine lights and coarse heavies cannot be

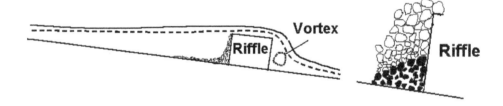

Figure 15.16 Effect of riffle over a deck of flowing film.

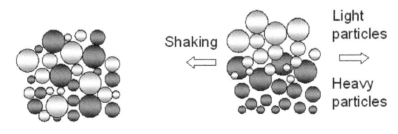

Figure 15.17 Stratification of particles due to the shaking motion.

separated as individual fractions (products) by tabling alone. Sorting classification followed by tabling gives complete separation. A comparison between the tabling alone and sorting classification followed by tabling is shown in Figure 15.18.

In Figure 15.18, R is the mixture of particles. A is the stratification after tabling operation as shown in Figure 15.17. B is the Stratification after sorting classification wherein fine lights are separated as overflow and coarse heavies are separated as first spigot product of a classifier, as shown in C. The middle fraction (second spigot product of a classifier), as shown in C, is subjected to the tabling operation. The resulting stratification is as in D. The stratified layers can then be separated as in E. With this example, it can be concluded that sorting before tabling as indicated in article 13.1.2 can result in complete separation. It is to be reiterated that this separation takes place due to the stratification in tabling being reverse to that of the stratification in classification.

A **Shaking table** is a device which utilizes reciprocating motion and cross riffles in addition to flowing film. The **Wilfley table** (Figure 15.19) is the most widely used shaking table. It consists of a slightly inclined deck on to which feed, at about 25% solids by weight, is introduced at the feed box and is distributed along the feed launder, wash water is distributed along the balance length of the top of the deck. The table is given a shaking motion longitudinally using a slow forward stroke and a rapid return, which causes the mineral particles to move along the deck parallel to the direction of motion. In this table, the riffles run parallel with the long axis and are tapered from the maximum height on the feed side (nearest the shaking mechanism) till they die out near the opposite side, part of which is left smooth. Where the riffles stand high, a certain amount of eddying movement occurs, aiding the stratification and jigging action between the riffles. As the load of material is jerked across the table, the uppermost layer is swept or rolled over the riffles. In this way the uppermost layer of

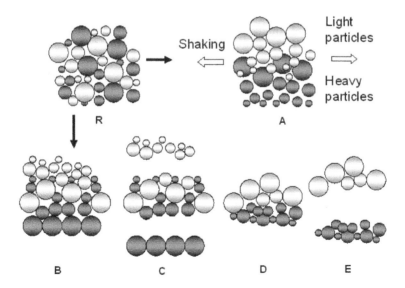

Figure 15.18 Comparison between tabling and sorting followed by tabling.

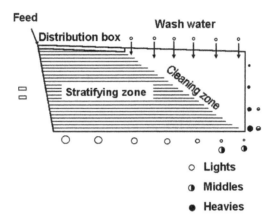

Figure 15.19 Wilfley table.

sand is repeatedly moved over riffle after riffle until it leaves the deck. As the water film is thinnest while climbing over the riffle, any suspended solids drop into the trough and move crosswise as a result of the mechanical shaking movement.

The high density particles settle into the troughs between the riffles and the motion of the table throws them toward the right side of the table and finally discharge them at the corner diagonally opposite to the feed corner. These particles are stratified while they travel. The light particles which remain in suspension are washed across the table by the water owing to the inclination of table. The other particles (middlings) discharge at intermediate places between lights and heavies. Tabling can be made a dry operation by injecting air upwards through the perforated deck in which case the operation is called **pneumatic tabling**.

Tables are successfully used for the recovery of heavy minerals from beach sands, chromite beneficiation and the like. These are normally operated on feed sizes in the range of 3 mm to 100 microns. The particles of size less than 100 microns are treated in **slimes tables** whose decks have a series of planes rather than riffles.

Vanners, GEC Duplex concentrator, Bartles cross-belt separator, Bartles-Mozley table are the other stirred bed flowing film concentrators. Knelson concentrator, Falcon Concentrator and Multi-Gravity Separator (MGS) are the Centrifugal separators. The Knelson and Falcon concentrators are high-speed centrifugal separators that set a centrifugal force to the particles in the slurry against a fluidisation water flow. The MGS combines the centrifugal motion of an angled rotating drum with the oscillating motion of a shaking table, to provide an enhanced gravity separation, particularly suited to fine particles.

Chapter 16

Froth flotation

In a beaker of water, if air is introduced from the bottom, air bubbles are produced and rise to the surface of the water as the density of an air bubble is much less than that of water (Figure 16.1 A). Similarly air bubbles also rise if air is introduced in a pulp containing mineral particles. If a mineral particle of high density adheres to the air bubble, the air bubble along with the mineral particle rises to the surface because the apparent density of the air bubble and the adhered mineral particle is less than that of water. If many mineral particles are adhered to the air bubble, still the air bubble rises to the surface as the apparent density of the air bubble and adhered mineral particles is less than that of water due to the relatively large volume of the air bubble (Figure 16.1 B). This concept is the basis for froth flotation operation.

To use this concept or phenomenon, the following are required:

1   Method to make the mineral particles adhere to the air bubble.
2   Method to keep the air bubble alive when it reaches the surface of water. It is known that the air bubble collapses when it reaches the surface of water. When it is collapsed, the adhering mineral particles are dropped into the water. So the air bubble must be kept alive on the surface of the water for a sufficient length of time for it to be removed from the surface.

The following are the methods adopted in practice:

1   To make the mineral particles adhere to the air bubble whilst it is rising through the pulp, the mineral particles are treated with suitable chemicals to acquire adhering properties.
2   To prevent the air bubble from collapse, some other chemicals are used to prevent the collapsing of the air bubble or in other words to increase the life of the air bubble.

The required mineral particles after adhering to the air bubble float to the surface along with the air bubble. The aggregation of several such mineral adhered air-bubbles forms the froth on the surface of the pulp (Figure 16.1 C). Hence this operation of mineral separation is named **Froth flotation**.

Flotation is a method of separating the minerals in a relatively finely divided state. It utilises the differences in physico-chemical surface properties of particles of various

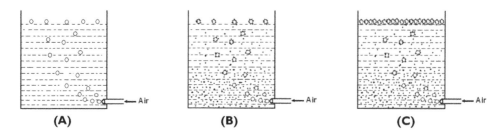

Figure 16.1 Process of rising air bubbles and forming froth.

minerals. This method can only be applied to relatively fine particles (less than 150 μm). If the particles are too large, the adhesion between the particle and the bubble is less than the weight of the particle and the bubble drops the mineral particles. The air bubbles can only stick to the mineral particles if they can displace water from the mineral surface. This can only happen if the mineral is water repellent or hydrophobic. Air bubbles, after reaching the surface, can continue to hold the mineral particles if they can form a stable froth. If not, air bubbles will burst and drop the mineral particles. In order to achieve the favourable conditions for froth flotation, the pulp is treated with various chemical reagents known as flotation reagents. The chemicals used for treating the mineral particles to make them to adhere to the air bubbles are called collectors and the chemicals used to increase the life of the air bubbles are called frothers.

The majority of the ores mined at present require fine grinding for a high degree of liberation of valuable minerals, and thus the flotation becomes the only possible means of beneficiation for higher grades and recoveries. Flotation treatment is extensively applied to the concentration of metalliferous minerals, both sulphides and oxides. About 90% of the world's important ores of lead, zinc and copper are upgraded by flotation operation. The flotation technique is also extensively used in upgrading of non-metallic ores like graphite, limestone, fluorite, fluorspar, clay, rock phosphate and coal.

16.1 CONTACT ANGLE

Contact angle (θ) is an angle of contact of an air bubble with the surface of a solid measured across the water (Figure 16.2). It is a convenient measure of the forces of adhesion between the bubble and the solid surface. The contact angle marks the position of equilibrium between the solid-water and water-air surfaces on a wetted surface i.e., it is the position of equilibrium between three tension forces; the surface tension of water T_{WA}, surface tension of solid mineral T_{MA} and interfacial tension T_{MW} between the solid mineral and water.

If the surface tension of the solid mineral T_{MA} is more, the water is pulled over the solid till an acute angle θ is reached, when the component of water tension T_{WA} together with the interfacial tension T_{MW} is sufficient to bring about equilibrium. Under these conditions, the solid shows a preference for water. If the interfacial tension T_{MW}

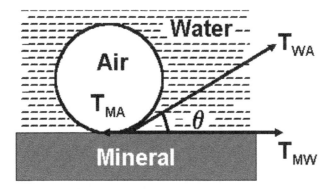

Figure 16.2 Contact angle.

is greater, the water will be drawn back and an obtuse angle will form. Under these conditions the solid has a preference for air.

When the solid shows affinity for water $T_{MA} = T_{MW} + T_{WA} \cos \theta$ 16.1.1

When the solid shows affinity for air $T_{MW} = T_{MA} + T_{WA} \cos \theta$ 16.1.2

Therefore, $T_{WA} \cos \theta$ is a measure of degree of wetting. When the contact angle is nil, $\theta = 0$, $\cos \theta = 1$, the degree of wetting is maximum. When the contact angle is 180°, $\theta = 180°$, $\cos \theta = -1$, the water will contract its extent and the degree of wetting is at a minimum. Since there is always some adhesion between solids and liquids in contact, there is no such thing as complete non-wettability, i.e., a contact angle of 180°. Adherence of the mineral particle to the air bubble depends on contact angle. As the contact angle increases, adherence increases and hence floatability increases. Minerals with high contact angle are called aerophilic (hydrophobic), i.e. they have higher affinity for air than for water. Most minerals are aerophobic (hydrophilic) in their natural state. To make the valuable mineral particles aerophilic, reagents called collectors are added to the pulp which adsorb on mineral surfaces, increases contact angle and facilitates bubble attachment. Many freshly formed mineral surfaces exhibit a natural contact angle of a few degrees. Graphite and some coals have high contact angles to float without aid of a collector. They are said to have natural floatability.

16.2 FLOTATION REAGENTS

Flotation reagents are substances added to the ore pulp prior to or during flotation in order to make it possible to float valuable mineral particles and not to float the gangue mineral particles. Important flotation reagents are collectors, frothers, depressants, activators and pH regulators.

A **Collector** is a chemical reagent, either an acid, base or salt, and is hetero-polar in nature; the polar part of it has an affinity towards a specific mineral and the non-polar part has an affinity towards an air bubble. A small amount of collector is added to the pulp and agitated long enough that the polar part is adsorbed on to the mineral to be floated whilst the non-polar part is oriented outwards and makes the surface of mineral particles hydrophobic. The collector increases the contact angle of the valuable mineral particles.

Collectors are broadly classified as anionic, cationic and oily collectors. Anionic and cationic collectors are ionizable organic compounds. They are said to be an anionic or cationic collector whether the ion that carries hydrocarbon group is anion or cation. Oily collectors are oily liquids which spread out thinly on solid surfaces to cause bubble attachment.

Xanthates, Dithiophosphates (known as Aerofloats), Dithiocarbamates, Fatty acids and soaps are the important anionic collectors. Xanthates are the most widely used collectors for flotation of sulphide minerals. The general formula of sodium or potassium xanthate is:

$$\text{ROC-S Na (K)}$$
$$\overset{\|}{\text{S}}$$

The R group, in the case of ethyl and isopropyl xanthates, are C_2H_5 and C_3H_7 respectively and is shown below:

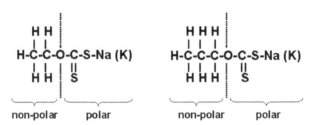

The general formulae for other anionic collectors are:

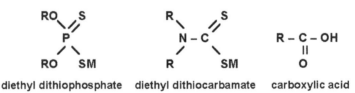

diethyl dithiophosphate　　diethyl dithiocarbamate　　carboxylic acid

Dithiophosphates and dithiocarbamates are used in the flotation of sulphide minerals; carboxylic collectors are used for flotation of non-sulphides and non-silicates.

Cationic collectors are used for oxide and silicate minerals including quartz. Amines are the most commonly used cationic collectors. Oily collectors normally used are petroleum products, blast furnace oils, coal-tar and wood-tar creosotes. They are used in flotation of oxidized metalliferous ores and gold ores.

A **Frother** is a chemical reagent and is heteropolar in nature; the polar part of it has an affinity for water and the non-polar part has an affinity for gas or repulsion for water. The frother acts upon the gas water interface. The addition of a frother decreases the surface tension of water and increases the life of bubbles produced.

The main objective of a frother is to permit the production of a sufficiently stable froth to hold the mineral particles that form a network around the bubbles until they are removed from the flotation unit.

As a result of the addition of a frother, the gas bubbles, formed under the surface of the water, are more or less completely lined with a monomolecular sheath of frother molecules which allows each bubble to come in contact with other bubbles. This forms a froth. Thus a froth is simply a collection of bubbles.

Cresylic acid and pine oil are the most widely used frothers. A wide range of synthetic frothers are now in use in many plants. Methyl Iso-Butyl Carbinol (MIBC) is most important among the synthetic frothers. The following are the chemical formulae of these organic reagents.

Cresylic Acid   Pine oil   Methyl Isobutyl Carbinol (MIBC)

Eucalyptus oil, camphor oil and sagebrush oil are used when they are more cheaply available than the common frothing agents.

To have an independent control, the frother should not have a collecting property. The reagents having both frothing and collecting properties are known as **frother-collectors.** The compounds like fatty acids, sulphonates and amines which are in use as collectors have also frothing properties. Kerosene is a frother-collector used in coal flotation.

Other chemical reagents, depressants, activators and pH regulators, called **modifiers,** are used extensively in flotation to modify the action of the collector, either by intensifying or reducing its water – repellent effect on the mineral surface. Thus they make collector action more selective towards certain minerals.

Depressants are inorganic chemicals. They react chemically with the mineral particle surfaces to produce insoluble protective coatings of a wettable nature making them non-floatable even in the presence of a proper collector. Thus formed protective coatings prevent the formation of collector film. Sodium or potassium cyanide is a powerful depressant for sphalerite and pyrite. A combination of sodium cyanide and zinc sulphate is more effective in depressing zinc sulphide minerals, sphalerite and marmatite. They also have a depressing action on pyrite. Lime is sometimes used to depress pyrite in sulphide flotation. Sodium or potassium dichromate is used to depress galena.

Sodium silicate is extensively used for the depression of silicates and quartz. Sodium silicate is much used as a **dispersant** for removing slimes from particle surfaces of sulphide minerals. Sulphuric acid is used to depress quartz in soap flotation. Lactic acid is a powerful depressant in iron sulphide flotation. Metaphosphates are used as depressants for non-silicates like barite, fluorite and calcite.

Activators, generally inorganic compounds, can modify the surface of non-floatable or poorly floatable mineral particles by adsorption on particle surface so that the collector may film the particle and induce flotation. An example of this is the use of copper sulphate in the flotation of sphalerite. Copper sulphate dissociates into

Table 16.1 Quantities of flotation reagents.

Reagent	Quantity in pounds per ton
Collectors	0.05 to 2.5
Frothers	0.025 to 0.25
Depressants	0.05 to 1.0
Activators	0.5 to 2.0
pH regulators	0.5 to 10.0
Dispersants	0.1 to 0.7

copper ions in solution and copper sulphide is formed at the surface of sphalerite. Then it reacts with xanthate and forms insoluble copper xanthate which makes the sphalerite surface hydrophobic. In the flotation of lead-zinc ore, after lead flotation, the sphalerite is activated with copper sulphate and floated. Copper sulphate also activates depressed pyrite when added in sufficient quantity.

Sodium sulphide is used to activate oxide minerals of lead, zinc and copper such as cerussite, smithsonite and malachite. As sodium sulphide imparts sulphide surface to the mineral particles to facilitate for collector coating, this activator is also known as **sulphidizer**. Sodium sulphide is also used to float previously depressed pyrite. It has dispersing and depressing effects on sulphide minerals when added in large quantities.

pH regulators are used to modify the alkalinity or acidity of a flotation circuit or in other words to control the pH of the pulp. The pH of the pulp has an important and sometimes very critical controlling effect on the action of the flotation reagents. Common pH regulators are lime and soda ash for creating alkaline conditions, sulphuric and hydrochloric acids for creating acidic conditions.

The quantity of reagents used in flotation varies from ore to ore and day to day or hour to hour for one ore. Small quantities of reagents are normally required. Low and high quantities of the reagents are of no use. In certain cases, increasing quantities results in other effects. For example, use of increased quantities of collector tends to float other mineral particles (other than required) also. Table 16.1 shows the approximate quantities of various reagents used in flotation.

16.3 TYPES OF FLOTATION

As a mineral is selected and floated in flotation, it is called **selective flotation**. As a required mineral is selected and floated, it is also called **direct flotation**. When an unwanted mineral is selected and floated, in which case, sink is the required product, it is called **reverse flotation**. When an ore contains two or more valuable minerals, one valuable mineral is selected and floated first and second valuable mineral is floated from the tailings. This flotation is called **differential flotation**. Alternatively, when all the valuable minerals are selected and floated, it is called **bulk flotation**. The individual minerals are selected and floated one after the other from the floats.

16.4 FLOTATION MACHINES

A flotation machine is the equipment used to carry out flotation operation. It provides the hydrodynamic and mechanical conditions which effect the separation. Basically the flotation machine must include:

1 Means for receiving the pulp.
2 Means for agitation and mixing the pulp.
3 Means for settling and discharging the pulp.
4 Means for air introduction and dispersion.
5 Means for discharging the froth.

The most commonly used flotation machines are of two types, namely mechanical type and pneumatic type. In a mechanical type flotation machine a mechanically driven impeller agitates the pulp and disperses the incoming air into small bubbles. The air may be drawn in by suction created by the impeller or may be introduced to the base of the impeller by an external blower. These flotation machines are often composed of several identical cells arranged in series in such a way that one cell receives the de-frothed pulp (tailing of preceding cell) as feed. Such a series of cells are called a bank. The Denver sub-aeration machine (Figure 16.3) is well known cell-to-cell machine.

The pulp from the weir of the preceding cell flows through the feed pipe on to the rotating impeller. The positive suction created by the impeller draws air through a hollow standpipe, sheared into fine bubbles by impeller and intimately mixed with the pulp. The diffuser arranged around the impeller prevents the agitation and swirling of the pulp above the impeller. The stationary hood above the impeller prevents the pulp

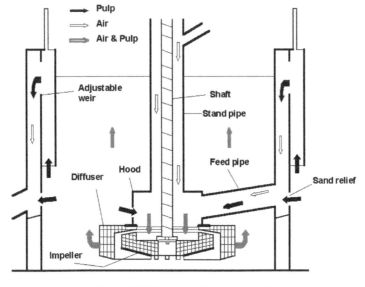

Figure 16.3 Denver sub-aeration cell.

in the cell having direct contact with the impeller. The mixture of bubbles and pulp leaves the impeller and bubbles ascend with their mineral load to form the froth. As the bubbles move to the pulp level, they are carried to the overflow lip by the crowding action of succeeding bubbles and removed by froth paddles. Pulp from the cell flows over the adjustable weir on to the impeller of the next cell. Particles which are too heavy to flow over the weir are by-passed through sand relief ports.

In an open-flow type machine, intermediate partitions and weirs between cells are eliminated. The pulp is free to flow through the machine without interference.

In pneumatic machines, air is introduced through the porous bottom of the cell. The air used in these machines not only creates aeration and produces froth but also maintains suspension by circulation.

16.5 FLOTATION OPERATION

The flotation operation is generally carried out in three stages namely roughing, scavenging and cleaning, called a flotation circuit (Figure 16.4). Each stage consists of a bank of cells and the number of cells in a bank is primarily depends upon the residence time of the pulp in the cells and the required throughput. The reagent conditioned feed pulp is treated in a first bank of cells called roughers. The tailing from the rougher cells, which may still contain some valuable mineral particles, is treated in another bank of cells called scavengers. The concentrate from the rougher cells is further treated in a bank of cells called cleaners to obtain high grade final concentrate. The scavenger concentrate and cleaner tailing are re-fed to the rougher bank to be treated with the fresh feed pulp. The scavenger tailing is the final tailing. It is to be noted that the pulp is treated in a conditioner, called **conditioning**, with necessary reagents like depressant and collector, prior to flotation in order to convert the mineral particles to respond readily in a flotation cell. Conditioning the pulp reduces its residence time in the flotation cell and hence the capacity of the cell increases.

16.6 COLUMN FLOTATION

The column flotation technique, which is a recent development, uses the principle of counter current wash-water flow for better separation particularly when operating

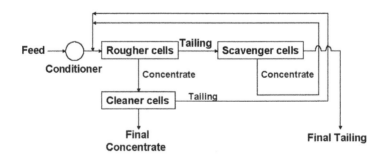

Figure 16.4 A typical flotation circuit.

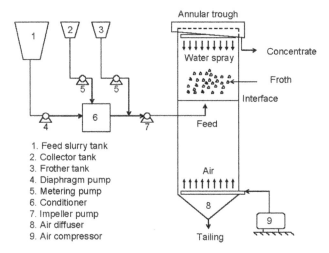

1. Feed slurry tank
2. Collector tank
3. Frother tank
4. Diaphragm pump
5. Metering pump
6. Conditioner
7. Impeller pump
8. Air diffuser
9. Air compressor

Figure 16.5 Flotation column.

on fine materials. The flotation column is a simplest form of pneumatic type flotation machine. It consists of a tall cylindrical column having the height to diameter ratio of more than 10 (Figure 16.5). The reagent conditioned feed pulp enters somewhat at the middle of the column. Compressed air is admitted through a distributor near the bottom of the column. In the section below the feed point, called recovery section, the feed pulp travels downwards against the rising air bubbles. Valuable mineral particles adhere to the air bubbles and are transported to the top part of the column, called washing section. Gangue mineral particles that are loosely attached to the bubbles are washed down by water sprays and only clean froth rises and discharged through annular trough.

The basic advantage of column flotation is the production of high grade concentrate without the loss of recovery. A single flotation column can replace five to six stages of operations involving conventional cells and yet achieve better performance. There are considerable savings in reagent requirement. The column occupies less floor space.

16.7   FLOTATION PRACTICE OF SULPHIDE ORES

Flotation operation was initially developed to treat the sulphides of lead, zinc and copper. In a typical flotation practice of lead-zinc ore, the different reagents used are sodium cyanide and zinc sulphate as depressants for pyrite and sphalerite, potassium ethyl xanthate as collector for lead circuit to float galena, sodium isopropyl xanthate in zinc circuit to float sphalerite, copper sulphate to activate already depressed sphalerite in zinc circuit, lime as pH regulator, and crysilic acid as frother in both circuits. In some plants, Methyl Iso-Butyl Carbinol (MIBC) is used as a frother. For the flotation of chalcopyrite (copper mineral in most of the copper concentrators), sodium isopropyl xanthate and pine oil are the collector and frother respectively in most of the plants. Soda ash is used as a pH regulator.

Chapter 17

Magnetic separation

Magnetic separation is a physical separation of particles based on the magnetic property of the mineral particles. The property of a material which determines its response to a magnetic field is the magnetic susceptibility.

17.1 MAGNETISM

According to the Weber's theory of magnetism, molecules of all substances are basically magnets in themselves, each having a north and south pole. In a magnet, these molecular magnets lie in line so that north poles of molecular magnets are on one end and south poles on the other. Thus, two definite poles N and S are developed at the ends.

Magnetic field (or force) of a magnet is the surrounding space through which its influence extends. The magnitude of force between two magnetic poles by coulomb's law is:

$$F = \frac{m_1 m_2}{4\pi\mu r^2}$$

17.1.1

where the force F is expressed in Newton's, m_1 and m_2 are the strengths of the poles expressed in Weber, r is the distance between the poles in meters, and μ is a constant depending on the medium.

$$F = \frac{m_1 m_2}{4\pi\mu_o \mu_r r^2} \quad \text{in a medium}$$

17.1.1a

$$F = \frac{m_1 m_2}{4\pi\mu_o r^2} \quad \text{in air (or vacuum or free space)}$$

17.1.1b

μ_o = **Absolute permeability** of vacuum or air, Henries/metre (H/m), weber/A-m
μ_r = **Relative permeability** of the medium surrounding the poles, dimensionless

The value of μ_o is $4\pi \times 10^{-7}$ H/m.
The value of μ_r is different for different media.
If, in the equation 17.1.1b, $m_1 = m_2 = m$ (say), $r = 1$ metre, $F = \frac{1}{4\pi\mu_o}$ newton.

Then, $m^2 = 1$ or $m = \pm 1$ weber (SI unit).

Hence, a **unit magnetic pole** may be defined as that pole which when placed in vacuum at a distance of one metre from a similar and equal pole repels it with a force of $1/4\pi\mu_o$ Newtons.

Magnetic field intensity or **Magnetic field strength** or **Magnetizing force** (H) at any point within a magnetic field is the force experienced by a N-pole of one Weber placed at that point. This magnetizing force induces the lines of force through a material. The unit of H in SI system is Ampere/metre (1 ampere/metre = $4\pi \times 10^{-7}$ tesla).

The total number of magnetic lines of force in a magnetic field is called **Magnetic flux.** It is represented by ϕ and the unit in SI system is weber. 1 weber = 10^8 lines or maxwells. **Magnetic flux density** (B) at any point is the magnetic flux passing per unit area at that point

$$B = \frac{\phi}{A} \text{ weber/m}^2 \text{ or tesla} \qquad\qquad 17.1.2$$

When an iron rod is placed in the magnetic field of intensity H, it becomes magnetized due to magnetic induction and the number of lines (called lines of induction) passing through iron is much greater. If B is the number of lines of induction per square meter in iron, then B is the **Magnetic induction** or **Magnetic flux density** developed in the rod. It is measured in Weber/m², Tesla, or Gauss (1 gauss = 10^{-4} tesla).

$$\therefore \text{ Permeability } \mu = \frac{\text{Magnetic Induction}}{\text{Impressed field intensity or Magnetising force}}$$

$$\mu = \frac{B}{H} \qquad\qquad 17.1.3$$

$$\Rightarrow \quad B = \mu H = \mu_o \mu_r H \text{ weber/m}^2 \qquad\qquad 17.1.4$$

For ferro-magnetic substances, such as iron, nickel, cobalt and alloys such as nickel-iron, and cobalt-iron, relative permeability is much greater than unity whereas for para-magnetic substances, such as aluminium, relative permeability is slightly greater than unity. For diamagnetic materials such as bismuth, the relative permeability is less than unity.

Magnetic flux density is the magnetic force at a point inside a magnetic material and is the resultant of the externally impressed magnetic field and the force due to the induced poles at the ends of the rod. If M is the strength of the induced magnetism, which is called as **magnetization** or **intensity of magnetization** and H is the impressed magnetic field, then:

Magnetic flux density $B = \mu_o(H + M)$ $\qquad\qquad 17.1.5$

In vacuum, $M = 0$, and it is extremely low in air.

$\therefore$ Magnetic flux density $B = \mu_o H$ $\qquad\qquad 17.1.6$

The **Magnetic susceptibility** of a material is defined as the ease with which a magnetic substance can be magnetized. **Magnetic susceptibility** S is the ratio of the intensity of magnetization produced in the material to the field strength which produces the magnetization.

$$\text{Magnetic susceptibility} = S = \frac{\text{Intensity of Magnetisation}}{\text{Impressed field intensity or Magnetising force}}$$

$$S = \frac{M}{H} \qquad\qquad 17.1.7$$

Magnetic flux density $B = \mu_o(H + M) = \mu_o(H + SH) = \mu_o(1 + S)H$

$1 + S = \mu_r$ = relative permeability
$\therefore$ Magnetic flux density $B = \mu_o\mu_rH = \mu H \qquad\qquad 17.1.4$

The field gradient is the rate at which the field intensity increases towards surface of the magnet. The force on the particle (F) by the magnet (the capacity of a magnet) to lift a particle is directly proportional to field intensity (H) and field gradient (dH/dl).

$$F \propto H\frac{dH}{dl} \qquad\qquad 17.1.8$$

An infinite number of combinations of field and gradient can be used to generate required lifting force. Production of a high field gradient as well as high intensity is therefore an important aspect in designing a magnetic separator.

17.2 ELECTRO-MAGNETISM

Magnetic field can be created by electric current. When an electric current (I) flows through a conductor, magnetic field is set up all along the length of the conductor. The magnetic lines of force are circular in a plane perpendicular to the current (Figure 17.1). The magnetic field near the conductor is stronger and becomes weaker as the distance increases from the conductor. The magnetic field becomes stronger if the current is increased and vice-versa. The direction of the field is reversed if the current is reversed.

If a wire is made into a loop and current is passed through it, the magnetic lines of force are circular around the wire all along its length. When a number of circular coils are arranged co-axially in such a way as same current flows in the same direction in the coils, then this arrangement is called a **solenoid**.

Solenoids are of two types (1) Iron-cored solenoid and (2) Air-cored solenoid. In the iron-cored solenoid (Figure 17.2 (a)), the set of circular coils are wound on an iron bar. In the air-cored solenoid (Figure 17.2 (b)), there is air inside the coils.

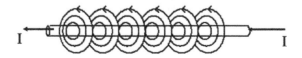

Figure 17.1 Magnetic lines of force.

Bars of soft wrought iron, soft steel or cast iron are magnetized by electric currents through the insulated copper or aluminum wire windings around the iron bars. These iron bars form electromagnets as long as the current flows through the windings. The poles of an electromagnet may be reversed by simply reversing the direction of flow of the electric current. The magnetism, or magnetic field, can be obtained of different intensities ranging from indefinitely weak to a certain maximum of strength. The intensity of the magnetic field depends on the size of the magnet, the form of it, the distance between the body to be attracted and the magnet, and the number of ampere turns in the magnet coil, that is, the product of the amperes of current flowing in the coil (I) times the number of turns around the core (N).

One important characteristic of magnetic flux is that each line is a closed loop. The complete closed path followed by group of lines of magnetic flux is known as magnetic circuit. Consider an iron ring wound with a number of turns as shown in Figure 17.3. When a current is passed through the winding, magnetic lines of force are set up in the ring. The path of these lines is called the magnetic circuit.

The magnetic pressure which sets up or tends to set up magnetic flux in a magnetic circuit is called **magneto-motive force** (m.m.f). m.m.f is produced by passing electric current through a wire of number of turns. It is measured in ampere turns, generally written as AT, and is the product of number of turns (N) and the current (I) flowing through these turns. m.m.f. = $N \times I$. The opposition to magnetic flux in a magnetic circuit is known as **reluctance**. The reluctance is directly proportional to the length of magnetic path (l), inversely proportional to the area of cross-section of the material through which flux is passing (A) and depends on the nature of the material.

$$\text{Reluctance} = \frac{l}{A\mu_o\mu_r} \qquad\qquad 17.2.1$$

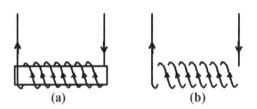

(a)   (b)

Figure 17.2 Types of solenoid.

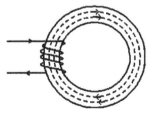

Figure 17.3 Magnetic circuit.

In any magnetic circuit, m.m.f. sets up magnetic flux and opposing to flux is offered by the reluctance of the magnetic path. Greater the m.m.f., greater is the flux i.e.

Flux $\propto$ m.m.f

$$\frac{\text{m.m.f}}{\text{flux}} = \text{constant} = \text{reluctance}$$

(or)   m.m.f = flux $\times$ reluctance

This is called ohm's law for magnetic circuit.

$$\text{Flux,} \quad \phi = \frac{\text{m.m.f}}{\text{reluctance}} = \frac{AT}{\frac{l}{A\mu_o\mu_r}} \qquad\qquad 17.2.2$$

Therefore Ampere-turns required $= AT = \phi \times \dfrac{l}{A\mu_o\mu_r}$ $\qquad\qquad$ 17.2.3

$$AT = \frac{\phi}{A} \times \frac{l}{\mu_o\mu_r} = \frac{B}{\mu_o\mu_r} \times l = H \times l \qquad\qquad 17.2.4$$

i.e. Ampere-turns required for any part of magnetic circuit = Field strength H in that part $\times$ length of that part

The practical magnetic circuits have mostly air gap i.e. magnetic flux is intended to pass through a certain distance in air as shown in Figure 17.4. Flux through air gap is called useful flux because it is only this flux which can be utilized for various purposes. It is, therefore, desired that total flux produced should pass through air gap. However, as air is not a perfect magnetic insulator, therefore, a part of the total flux returns by such paths as 'a' and 'b' and is called leakage flux. It is so called because it does not pass through the air gap rather it follows a leaking path.

In an electromagnet, the induction for a current of given intensity decreases rapidly with an increase in the air gap and the force of attraction decreases even more rapidly with an increase in the air gap. To get a suitable field at the lowest operating cost, it is recommended to reduce the air gap to a minimum, to reduce the reluctance to a minimum by using a highly permeable core, a large cross section and a short length and to use many turns and a small current.

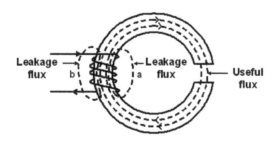

Figure 17.4 Magnetic circuit with air gap.

17.3   TYPES OF MINERALS

Based on magnetic susceptibility, minerals are classified into three broad groups as Ferromagnetic, Paramagnetic and Diamagnetic.

Ferromagnetic minerals have a very high susceptibility to magnetic forces and retain some magnetism when removed from the magnetic field. They are strongly attracted by the magnetic field. They can be concentrated by low-intensity magnetic separators. The principal ferromagnetic mineral separated is magnetite (Fe_3O_4).

Paramagnetic minerals are weakly attracted by magnetic field. They can be concentrated by high-intensity magnetic separators. Examples of paramagnetic minerals are ilmenite ($FeOTiO_2$), rutile (TiO_2), wolframite ($(Fe,Mn)WO_4$), hematite (Fe_2O_3), siderite ($FeCO_3$), chromite ($FeOCr_2O_3$), pyrrhotite (FeS) and manganese minerals. Hematite and siderite can be roasted to produce magnetite and can be separated by low-intensity magnetic separators.

Diamagnetic minerals are repelled along the lines of magnetic force to a point where the field intensity is smaller. The forces involved here are very small. They cannot be concentrated magnetically. Examples are quartz (SiO_2) and feldspar ($KAlSi_3O_8$ or $NaAlSi_3O_8$).

17.4   MAGNETIC SEPARATORS

The machines used for magnetic separation are called magnetic separators. The principle operation of a magnetic separator is:

> When a stream of ore particles is continuously passed through a field of magnetic force, the magnetic particles are attracted towards the source of magnetic force while the non-magnetics travel unaffected.

Magnetic separation, in an industry, is achieved by simultaneously applying a magnetic force on all particles of an ore which acts on magnetic particles and a combination of forces which acts in a different direction and affects both magnetic and nonmagnetic particles. The most commonly applied nonmagnetic forces are gravitational, centrifugal, and fluid drag. Other forces which usually enter in an incidental manner are frictional, inertial and attractive or repulsive inter-particle forces.

These forces depend on both the nature of the feed material and the character of the separator. The nature of the feed material includes its size distribution, magnetic susceptibility and other physical and chemical properties that may effects the various forces involved.

Magnets used in magnetic separators are of two types; Permanent magnets and Electromagnets. Permanent magnets are magnetized steel bars which retain magnetism indefinitely. Special-permanent magnet alloys produce a magnetic field at a constant level indefinitely after initial charging. The newer ceramic permanent magnets in the form of barium and strontium ferrites, and rare-earth-cobalt permanent magnet alloys are also being in use. Bars of soft wrought iron, soft steel or cast iron are magnetized by electric currents through the insulated copper or aluminum wire windings around the iron bars. These iron bars form electromagnets as long as the current flows through the windings.

Magnetic separators, can be classified into two groups based on their function:

1   **Tramp iron magnetic separators** are those used to remove tramp iron and protect the handling and processing equipment.
2   **Concentrators** are those used to separate the bulk of magnetic materials from a stream of ore. Magnetic separators that are used to remove small quantities of deleterious magnetic material from a product (e.g. from china clay) are termed as **purifiers**. The concentrators that remove magnetic material from a product and return to the process are termed as **reclaimers** (e.g. separation of medium particles such as magnetite, ferrosilicon etc. in Heavy Medium Separation process).

17.4.1   Tramp iron magnetic separators

Iron coarser than 3 mm is usually called tramp iron. Most commonly applied magnetic separators for removing tramp iron are suspended magnets, magnetic pulleys, and grate magnets. Suspended magnets are fixed position electromagnets of rectangular shape and installed over a conveyor belt or head pulley (Figure 17.5). Tramp iron attracted by the magnet is periodically removed. Cross-belt models (Figure 17.5) are used when the continuous removal of tramp iron is required. In a magnetic pulley, magnets are located within the head pulley of a belt conveyor (Figure 17.5). The tramp iron is attracted by the magnetic pulley and carried to distant place and discharged when the material comes out of magnetic field while the required material is discharged at the same place. Grate magnet consists of a series of magnetized bars (Figure 17.5) and is used to remove fine iron as well as tramp iron. Feed is usually vertical through the grate. These magnets must be periodically cleaned.

17.4.2   Concentrators

These magnetic separators are classified into dry and wet separators which may further classified into low- and high-intensity separators. Wet separators are generally used for the particles of below 0.5 cm size.

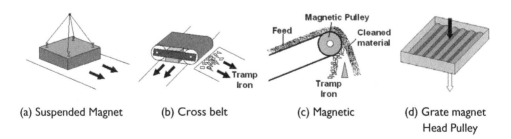

(a) Suspended Magnet   (b) Cross belt   (c) Magnetic   (d) Grate magnet Head Pulley

Figure 17.5 Tramp iron magnetic separators.

17.4.2.1   Dry magnetic separators

Drum separators are common low-intensity dry magnetic separators. These are applied in concentration and purification where magnetic particles to be removed are large in size and the magnetic responsiveness of the magnetic particles is high. This operation of picking up strongly magnetic particles at a coarser size is called **cobbing** operation. Drum separators consist of a rotating non-magnetic drum containing either permanent or electromagnets. Magnetic particles are attracted by the magnets, pinned to the drum, conveyed out of the field and discharged while the nonmagnetic particles are discharged unaffected. The magnetic drum is normally fed at the top vertical center, but, with appropriate magnet positioning, it can be fed at any convenient point or even front-fed or underfed (Figure 17.6).

Magnetic drum is also used for tramp iron removal. Magnetic head pulley and grate magnet used for tramp iron removal are also used as concentrators. Grate magnet is especially useful for purification of the product when it contains small amounts of unwanted magnetic particles.

Induced roll separator and cross-belt separator are the two high intensity dry magnetic separators. **Induced roll separator** (Figure 17.7) consists of a horseshoe magnet faced by an iron bridge bar or keeper and of two rolls, one opposite each pole. The magnetic circuit is completely in iron except for the very small clearance between the rolls and the bridge bar and for the gaps between the rolls and primary poles of horseshoe magnet. The rolls consist of alternate laminae of permeable and impermeable material with a serrated profile and produce the high field intensity and gradient required. The field strengths of up to 2.27 T are attainable in the gap between feed pole and roll. Non-magnetic particles are thrown off the roll into the tailings compartment, whereas magnetics are gripped, carried out of the influence of the field and discharged. The gap between the feed pole and rotor is decreased from pole to pole successively to separate weaker magnetic products.

Cross-belt separator consists of two or more horseshoe electromagnets arranged with poles of opposite sign in apposition, one electromagnet being below and the other above the feed belt. The poles of the upper magnets are wedge-shaped while the lower poles are flat. This concentrates the field and attracts the paramagnetic minerals

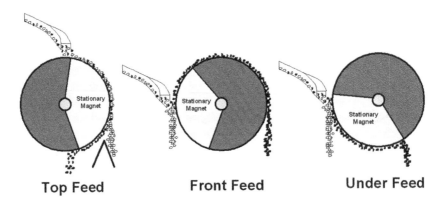

Figure 17.6 Dry magnetic drum separators.

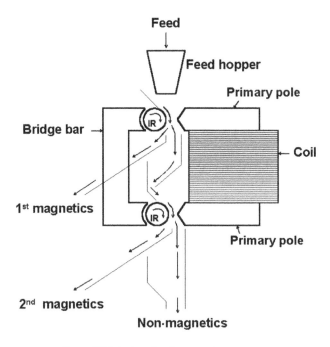

Figure 17.7 Induced roll magnetic separator.

toward the wedge-shaped poles. Dry material is fed in a uniform layer on to the conveyor belt and is carried between the poles. The cross-belt prevents the magnetic particles from adhering to the poles and carries them out of the field.

Disc separator is the modification of the cross belt separator wherein discs revolve above a conveyor belt and are magnetized by induction from powerful stationary electromagnets placed below the belt. A disc separator permits a much smaller air gap and provides a greater degree of selectivity in separating the minerals differing in their magnetic susceptibility slightly.

17.4.2.2 Wet magnetic separators

Drum separators are most common low-intensity wet magnetic separators. They consist of a rotating non-magnetic drum containing three to six stationary magnets, either permanent or electromagnets, of alternating polarity. Magnetic particles are lifted by the magnets and pinned to the drum and are conveyed out of the field leaving the gangue particles. Water is introduced into the machine to provide a current which keeps the pulp in suspension. Three types of drum separators namely **concurrent** type, **counter-rotation** type and **counter-current** type (Figure 17.8) are in use. Concurrent type separators are used when clean concentrate is required from relatively coarse material such as heavy medium recovery systems. Counter-rotation type separators are used in the roughing operation of larger tonnages. Counter-current type separators are designed for finishing operation on relatively fine material.

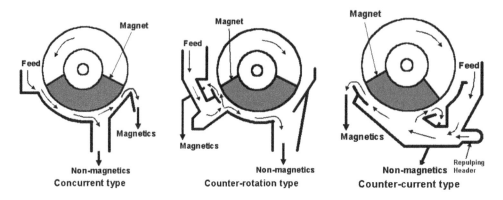

Figure 17.8 Low-intensity wet drum magnetic separators.

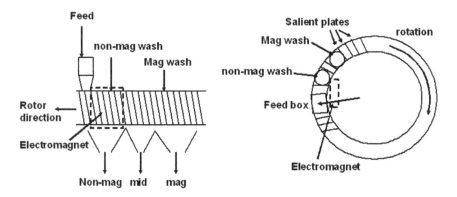

Figure 17.9 Wet high-intensity magnetic separator.

The Readings are wet high intensity magnetic separators (WHIMS). The unit (Figure 17.9) consists of a rotating carousel, which has vertically inclined salient plates through which feed slurry is passed. As the carousel rotates, it passes through fields of magnetic influence generated by surrounding electromagnets, followed by fields of no magnetic influence. The magnetic grains are initially held up in the plates while the non magnetic grains are washed through into a launder below. When the plates are in the non magnetic field the magnetic grains are washed off into a separate launder below.

In order to attract and hold paramagnetic particles, both high magnetic field and a high field gradient are required. These high gradient magnetic separators (HGMS) produce field strength of about 2 T. In HGMS, the slurry is passed through a container where it is subjected to a high-intensity, high gradient magnetic field. The container is packed with a **capture** matrix made of stainless steel wool. In the presence of magnetic field, paramagnetic particles become magnetized and are trapped in the matrix while the unaffected non-magnetic particles pass through the container. When the matrix is loaded to its magnetic capacity, the slurry feed is stopped and the electric power is cut off. The matrix is then backwashed to remove magnetic particles. Afterwards, the feed and the power are resumed and the entire process is repeated.

In the continuous separator, the capture matrix is a continuous, segmented metal belt. The slurry is introduced while capture matrix passes through the magnetic section of the separator. The magnetized particles are trapped in the capture matrix. The belt then travels in to the washing section to wash magnetic particles trapped on the belt. The cleaned belt section then returns for the entire process to be repeated.

Chapter 18

Electrical separation

Electrical separation is a physical separation of particles based on the electrical properties of the mineral particles. The following are the three important electrical separation processes:

1 Electrostatic separation.
2 Electrodynamic (High Tension) separation.
3 Dielectric separation.

Electrostatic and Electrodynamic separation processes utilize the difference in electrical conductivity between various minerals in the ore feed whereas dielectric separation utilizes the differences in dielectric constant of mineral particles.

Electrical conductivity is defined as current density (flow of electric charge per unit area of cross section) per unit applied electric field. It is the reciprocal of the resistivity or specific resistance of a conductor. It is measured in ohm^{-1} cm^{-1}.

The conductivity of minerals varies enormously from one mineral to another. Native metals like Au, Ag, Cu, etc., many metal sulphides like Chalcopyrite, Galena, Molybdenite, etc., and some other minerals like Graphite, Garnet, Chalcocite, etc., are good conductors, whereas most gangue and non-metallic minerals like Calcite, Gypsum, Monazite, etc., are relatively poor conductors of electricity.

Knowledge of charge and charge interactions is helpful in understanding the phenomenon involved.

18.1 CHARGE AND CHARGE INTERACTIONS

The atom consists of a **nucleus** surrounded by negatively charged electrons. The proton and neutron are located in the nucleus. The proton is charged positively. The neutron does not possess a charge and is neutral. The amount of charge on a single proton is equal to the amount of charge on a single electron. If an atom contains equal numbers of protons and electrons, the atom is described as **electrically neutral**. Any particle, whether an atom, molecule or ion, which contains less electrons than protons is said to be **positively-charged**. Conversely, any particle which contains more electrons than protons is said to be **negatively charged**.

It can be said that positive electrification of a body results from deficiency of the electrons whereas negative electrificiation of a body results from an excess of electrons.

18.1.1   Conductors and insulators

Conductors are materials which permit electrons to flow freely from atom to atom and molecule to molecule. An object made of a conducting material will permit charge to be transferred across the entire surface of the object. If a charge is transferred to the object at a given location, that charge is quickly distributed across the entire surface of the object. If a charged conductor is touched to another object, the conductor can even transfer its charge to that object. The transfer of charge between objects occurs more readily if the second object is made of a conducting material. Conductors allow for charge transfer through the free movement of electrons.

Insulators are materials which impede the free flow of electrons from atom to atom and molecule to molecule. If a charge is transferred to an insulator, the excess charge will remain at the initial location of charging. The particles of the insulator do not permit the free flow of electrons. Conductive objects are often mounted upon insulating objects to prevent transfer of charge from the conductive object to its surroundings.

Examples of conductors include metals, aqueous solutions of salts, graphite, water and the human body. Examples of insulators include plastics, paper, rubber, glass, wood and dry air. Some materials are neither good conductors nor good insulators, since their electrical characteristics fall between those of conductors and insulators. These in-between materials are classified as **Semi-conductors**. Germanium and silicon are two common semiconductors. Different materials may be placed along a continuum as in figure 18.1. The conductivity of a metal might be as much as a million trillion times greater than that of glass.

18.1.2   Polarization

Polarization is the process of separating opposite charges within an object. The positive charge becomes separated from the negative charge. By inducing the movement of electrons within an object, one side of the object is left with an excess of positive charge and the other side of the object is left with an excess of negative charge. Charge becomes separated into opposites. The polarization process always involves the use of a charged object to induce electron movement or electron rearrangement.

Polarization is not charging. When an object becomes polarized, there is simply a redistribution of the centers of positive and negative charges within the object.

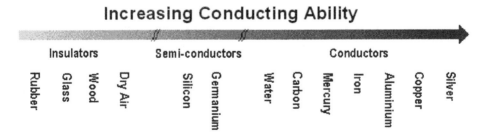

Figure 18.1 Classification of materials based on conducting ability.

Either by the movement of electrons across the surface of the object (as is the case in conductors) or through the distortion of electron clouds (as is the case in insulators), the centers of positive and negative charges become separated from each other. The atoms at one location on the object possess more protons than electrons and the atoms at another location have more electrons than protons.

18.2 METHODS OF CHARGING

1. Charging by Friction: The presence of different atoms in objects provide different objects with different electrical properties. One such property is known as **electron affinity**. The frictional charging process results in a transfer of electrons between the two objects which are rubbed together. Rubber has a much greater attraction for electrons than animal fur. As a result, the atoms of rubber pull electrons from the atoms of animal fur, leaving both objects with an imbalance of charge. The rubber balloon has an excess of electrons and hence is charged negatively. The animal fur has a shortage of electrons and is charged positively.

As mentioned, different materials have different affinities for electrons. By rubbing a variety of materials against each other and testing their resulting interaction with objects of known charge, the tested materials can be ordered according to their affinity for electrons. Such an ordering of substances is known as a **triboelectric series**.

2. Charging by Induction: The phenomenon of an uncharged body getting charged merely by the nearness of a charged body is known as induction. Induction charging is a method used to charge an object without actually touching the object to any other charged object.

Charging a Two-Sphere System: Two metal spheres are supported by insulating stands so that any charge acquired by the spheres cannot travel to the ground. The spheres are placed side by side (Figure 18.2 (a)) so as to form a two-sphere system. If a rubber balloon is charged negatively and brought near the spheres, electrons within the two-sphere system will be induced to move away from the balloon. Being charged negatively, the electrons are repelled by the negatively charged balloon. They are free to move about the surface of the conductor. Subsequently, there is a mass migration of electrons from sphere A to sphere B. This electron migration causes the two-sphere system to be polarized (Figure 18.2 (b)). Sphere A has an overall positive charge and sphere B has an overall negative charge.

Once the two-sphere system is polarized, sphere B is physically separated from sphere A using the insulating stand. Then the negative charge likely redistributes itself uniformly

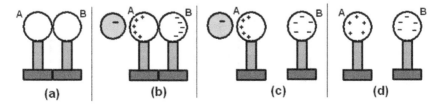

Figure 18.2 Charging by induction with negatively charged object.

about sphere B (Figure 18.2 (c)). Meanwhile, the excess positive charge on sphere A remains located near the negatively charged balloon. As the balloon is pulled away, there is a uniform distribution of charge about the surface of both spheres (Figure 18.2 (d)). This distribution occurs as the remaining electrons in sphere A move across the surface of the sphere until the excess positive charge is uniformly distributed.

When a positively charged balloon is brought near Sphere A, the positive charge induces a mass migration of electrons from sphere B towards (and into) sphere A. Negatively charged electrons throughout the two-sphere system are attracted to the positively charged balloon. This movement of electrons from sphere B to sphere A leaves sphere B with an overall positive charge and sphere A with an overall negative charge (Figure 18.3 (f)). The two-sphere system has been polarized. With the positively charged balloon still held nearby, sphere B is physically separated from sphere A. The excess positive charge is uniformly distributed across the surface of sphere B. The excess negative charge on sphere A remains crowded towards the left side of the sphere, positioning itself close to the balloon. Once the balloon is removed, electrons redistribute themselves about sphere A until the excess negative charge is evenly distributed across the surface. In the end, sphere A becomes charged negatively and sphere B becomes charged positively (Figure 18.3 (h)).

Charging a Single sphere: When a negatively charged rubber balloon B is brought near a single sphere, negative electrons within the metal sphere will be repelled by the negatively charged balloon. There will be a mass migration of electrons from the left side of the sphere to the right side of the sphere causing charge within the sphere to become polarized (Figure 18.4 (b)). Once charge within the sphere has become polarized, the sphere is grounded.

The grounding of the sphere allows electrons to exit from the sphere and move to the earth (Figure 18.4 (c)) and the sphere acquires a positive charge (Figure 18.4 (d)). Once the balloon is moved away from the sphere, the excess positive charge redistributes

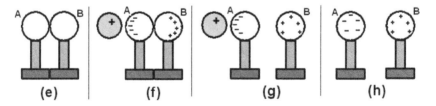

Figure 18.3 Charging by induction with positively charged object.

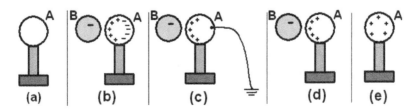

Figure 18.4 Charging single sphere by induction.

itself (by the movement of remaining electrons) such that the positive charge is uniformly distributed about the sphere's surface. Positive and negative charges of A are known as induced charges. The positive charge is called bound charge because it must remain on A so long as negative charge of B remains there. However the negative charge on the farther end of A is called free charge. If B is removed, then this positive charge will also go to earth, leaving A uncharged.

The fundamental principles of induction charging are:

- The charged object is never touched to the object being charged by induction.
- The charged object does not transfer electrons to or receive electrons from the object being charged.
- The charged object serves to polarize the object being charged.
- The object being charged is grounded; electrons are transferred between the ground and the object being charged (either into the object or out of it).
- The object being charged ultimately receives a charge that is opposite that of the charged object which is used to polarize it.

3. Charging by Conduction: Charging by conduction involves the contact of a charged object to a neutral object. If a charged object is touched to a neutral object, the neutral object becomes charged. It is often called **charging by contact**.

When a negatively charged object is made in contact with a neutral object, electrons move from the negatively charged object onto the neutral object. When finished, both objects were negatively charged. In case of charging with positively charged object, electrons move from neutral object to charged object. Having lost electrons to the positively charged body, there is a shortage of electrons on the neutral body and hence it is positively charged. The charged body is still charged positively, only it now has less excess positive charge than it had before the charging process began.

18.3   ELECTROSTATIC SEPARATION

If an ore contains conducting as well as non-conducting minerals/particles, electrostatic separation or high tension separation can be employed. Theoretically, it is not necessary that one of two minerals should be a good conductor and the other be a poor conductor, but the difference in their conductivity will affect the separation.

The basis of any electrostatic separation is the interaction between an external electric field and the electric charges acquired by the various particles. Particles can be charged by:

1   Contacting dissimilar particles.
2   Conductive induction.
3   Ion bombardment.

In every separation, two or more charging mechanisms occur.

In charging by contacting dissimilar particles, particles placed on surface are made to repeatedly contact one another as well as the surface; the surface will acquire electrons from one type of particles and give electrons to another type so that two types of particles are charged with opposite charges. When they are passed through an electrostatic separator consisting of two conducting plates across which a high voltage is applied, positively charged particles are attracted towards a negative plate and negatively charged particles are attracted towards a positive plate. This mechanism is utilized in free fall electrostatic separators. However, this is not the major mechanism in any of the electrostatic separators.

In charging by conductive induction, the particles are placed on a ground conductor in the presence of an electric field. Then the particles will rapidly develop a surface charge by induction. Both conducting and non-conducting particles will become polarized, but conducting particles will have an equi-potential surface through its contact with the grounded conductor. The non-conducting particle will remain polarized.

In charging by ion bombardment, corona discharge is obtained by appropriate shaping of the electrodes. Corona discharge is an electrical discharge which occurs when one of the two conducting surfaces (such as electrodes) of differing voltages have a pointed shape. A highly concentrated electric field, set up at the tip of the pointed shape electrode, ionizes the air (or other gas) around it. These ions charge the particles by bombardment when the mineral particles are caused to pass within the corona. For the discharge to a large diameter cylindrical surface, a fine wire parallel to the cylinder gives the optimum corona discharge.

The attraction of particles carrying one kind of charge (positive or negative) towards an electrode of the opposite charge is known as **lifting effect** as the particles are lifted towards the charged electrode. Mineral particles having a tendency to become charged with a definite polarity may be separated from each other by the use of lifting effect even though their conductivities may be very similar. For example, quartz is readily negatively charged and can be separated from other non-conductors by using an electrode carrying a positive charge. However, pure electrostatic separation (pure lifting effect) is relatively inefficient.

The other effect which makes the electrostatic separation more effective is the **pinning effect**. Here, non-conducting mineral particles, having received a surface charge from the electrode, retain this charge and are pinned to the earthed surface due to the attraction between the charged non-conducting particle and the grounded surface.

18.4   ELECTROSTATIC SEPARATOR

In an electrostatic separator (Figure 18.5), a large single electrode produces a strong electric field. When a particle is placed on the grounded rotor in the region of the electrostatic field influence, the particle rapidly develops a surface charge by induction. Whether the particle is a conductor or non-conductor, it will be polarized. However, a conducting particle rapidly becomes an equi-potential surface and has the same potential as the ground rotor. Therefore it is attracted towards the electrode, drawn away from the surface, and falls by gravity. The non-conducting particle continues to adhere to the rotor until it falls by gravity.

18.5   HIGH TENSION SEPARATOR

In a high tension separator (Figure 18.6), the mixture of ore minerals is fed on to a grounded rotor into the field of a charged ionizing electrode. The electrode assembly,

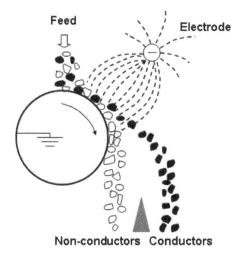

Figure 18.5 Electrostatic separator.

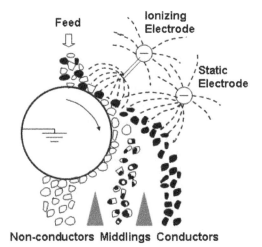

Figure 18.6 High tension separator.

consists of a lengthy fine wire supported by a brass tube, is supplied with DC supply of negative polarity. Then the ionization of air takes place which can be seen as corona discharge. This gives a high surface charge to the poor conducting mineral particles. These particles are attracted and pinned to the rotor. High conducting mineral particles are not charged as the charge is dissipated through the particles to the earthed rotor. Then they come under the influence of the electrostatic field of the non-ionizing electrode and are attracted (by the lifting effect) from the rotor surface. They are discharged away from the rotor. The charge of poor conducting mineral particles is slowly lost as the rotor rotates and the particles drop from the rotor as middlings

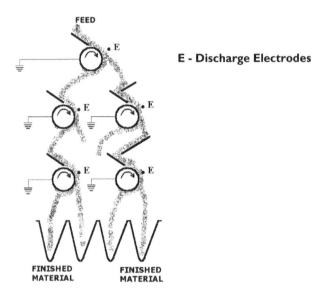

Figure 18.7 Multi-pass electrostatic separator.

or non-conductor product according to the intensity of the surface charge. Thus a combined effect of pinning and lifting is experienced in high tension separators. These separators operate on feeds of 60 to 500 microns size.

Since almost all minerals show some difference in conductivity, electrostatic separation is the universal concentrating method. It is widely used in beach sand beneficiation. The feed to electrostatic separation must be perfectly dry. For efficient operation, the feed should be in a layer, one particle deep, which severely restricts the throughput. Electrostatic separation is applicable to a limited size range, normally 20 mesh to 250 mesh. As complete separation is not likely to be obtained in a single pass electrostatic separator, it is usual practice to re-pass separated fractions through other separators as shown in Figure 18.7.

In Dielectric separation, ore consists of minerals of various dielectric constants is suspended in a non-conducting fluid whose dielectric constant is intermediate between that of two groups of minerals and a converging electric field is set up within the suspension. Then the particles having dielectric constant higher than the fluid travel in the direction of most rapid increase in electric field and the particles having dielectric constant lower than the fluid travel in the opposite direction. The dielectric separation process is applicable to particles finer than 60-mesh and is used as a research tool for study of minerals. It has no industrial application.

Chapter 19

Dewatering

Dewatering is the separation of pulp into two parts, one is relatively solid-free and the other is relatively liquid-free, with respect to original pulp. As most of the mineral beneficiation operations are conducted by the use of substantial quantities of water, the products are obtained in the form of pulp. Hence water has to be removed from the pulp to get the final product in dry form. The removal of water from the pulp is called dewatering. If the solid particles in the pulp are relatively coarser, screening the pulp results in removal of water by passing the water through apertures and retaining the solid particles on the screen. Draining is a fairly effective method to remove water from coarse sands, but if fine sands and slimes are present, they tend to run off with the water.

The pulps containing fine sands and slimes, flotation pulps for example, are dewatered usually in the following three stages:

1 Sedimentation.
2 Filtration.
3 Drying.

Sedimentation means gravity settling or subsidence of solids suspended in liquid. This operation, coupled with continuous overflow of water and withdrawal of partially dewatered solids (thick pulp) from the bottom, is called thickening. Much of the water is first removed by the thickening operation and the thick pulp so obtained is then filtered to produce moist filter cake. The filter cake, in many cases, is subjected to further processing directly. When it is required to eliminate the moisture, the filter cake is dried thermally to about 95% solids by weight. Removal of water by drying is the most expensive operation. Hence it is usual practice to eliminate as much water as possible by filtration, leaving only the moisture contained in the filter cake for removal by thermal drying.

19.1 THICKENING

Thickening is an operation of concentrating a relatively dilute slimy pulp into a thick pulp by allowing solid particles to settle under the influence of gravity force. Thickening is also considered as a classification with a difference that in thickening all the solid particles are allowed to settle whereas in classification only certain solid particles are allowed to settle.

Thickening is a complex phenomenon to analyse and its rate is difficult to predict as a large number of factors are involved. As the process of sedimentation (or thickening) proceeds, the slurry becomes more and more concentrated and as a result, the rate of settling of particles decreases. Sedimentation is thus a time-dependent phenomenon and its rate decreases with time. A preliminary study of the phenomenon, known as batch sedimentation test (Figure 19.1), is usually performed in the laboratory on a sample of the slurry. A sample of the slurry is taken in a graduated cylinder and is kept under observation. At the outset, the particles settle at their maximum hindered settling velocity and the rate of sedimentation will be essentially constant. Soon a zone of clear liquid (A) develops at the top, below it appears the thickening zone (B) of essentially uniform concentration of solids more or less similar to that of the original slurry and the heavy sludge or sediment that accumulates at the bottom constitutes the compression zone (D). A transition zone (C) may also appear between the thickening zone and compression zone. As settling proceeds, the interface between the thickening zone and the clear liquid zone descends down, decreasing the height of the thickening zone and increasing that of the clear liquid and compression zones. As the process is in progress, the density and viscosity of the suspension increases and the sedimentation rate gradually decreases. It reaches a critical point where the thickening zone (B) and the transition zone (C) disappears and only two zones, thickening and compression zones, exists. Further sedimentation results in compaction of the compression zone D by oozing out some liquid into clear liquid zone due to the weight of the clear liquid in zone A. This process of removal of some liquid from compression zone due to weight of clear liquid is known as **exudation**. The exudation process continues till the ultimate height of the compression zone is reached.

As very fine particles of a few microns in size settle extremely slowly, the thickening operation is very slow and cannot produce clarified liquid. In order to increase the settling rate, these fine solid particles are agglomerated or flocculated. **Flocculation** is defined as clustering, coagulation or coalescence of fine particles to form floccules or flocs in a liquid medium. Flocculation may be achieved by the addition of certain

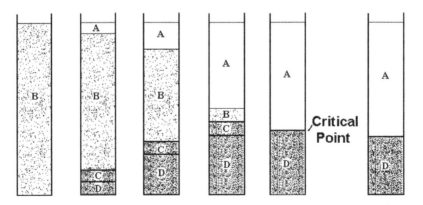

Figure 19.1 Batch sedimentation test.

reagents known as flocculating agents or flocculants such as lime, starch, glue, gelatine, alum, gypsum, sulphuric acid, copper sulphate etc. to the pulp. Molecules of the flocculant act as the bridges between separate suspended fine particles and form the flocs. These flocs settle rapidly leaving the clear liquid at the top.

The thickening operation is usually carried out in large tanks called **thickeners**. Thickener consists of a large cylindrical tank of 3 to 30 metres in diameter and 2.5 to 3 metres in depth with a very short conical bottom. Dorr thickener is one of such conventional thickeners and is shown in Figure 19.2. The feed is introduced continuously through a central semi-submerged feed well at a rate that allows the solid particles to settle at a safe distance below the overflow level. Thus clear water overflows into the top peripheral launder. The settled solids are swept by a slowly revolving raking mechanism attached to the central rotating shaft and positioned slightly above the tank bottom. Slow agitation of the slurry helps in reducing the apparent viscosity of the suspension. The sludge swept by the rakes is directed towards the center to facilitate easy removal by a suitable pump such as diaphragm pump.

In a thickener, the surface area must be large enough that the upward velocity of liquid is always lower than the settling velocity of the slowest-settling particle which is to be recovered. The degree of thickening produced is controlled by the residence time of the particles and hence by the thickener depth. Hence the diameter of the thickener is usually large compared with the depth. The diameter of the thickener is reduced by using Multi-tray thickener where a series of unit thickeners mounted vertically above one another. They operate as separate units, but a common central shaft is utilized to drive the sets of rakes. The high rate thickener and lamella thickener are the two thickeners used to save the floor space at the same time as increasing the settling rate.

In the thickening operation, the feed pulp containing about 15–30% solids is thickened to a pulp of 55–65% solids. The main aim of thickening is to obtain thick

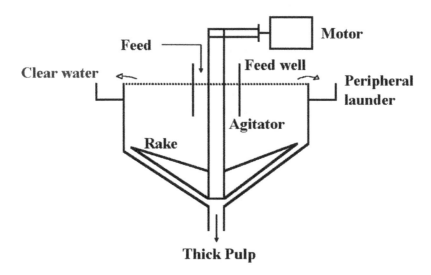

Figure 19.2 Dorr thickener.

pulp. If it is desired to obtain clear liquid rather than the thick pulp, the thickening operation is called **clarification**. This operation involves a dilute pulp of 1–5% solids. The unit used for clarification is similar to that of the thickener in design but called as **clarifier**. However, the clarifier cannot achieve 100% removal of solids in a reasonable length of time; although some operations in chemical plants produce overflows containing 3–4 ppm of solids. The treatment of tailing is done to obtain clarified water for recirculation.

Thickening can also be performed in a centrifugal field of force either by hydrocyclones or centrifuges. Thickening by hydrocyclone is simple and cheap but less efficient at very fine particle sizes.

19.2   FILTRATION

Filtration is the separation of finely divided solid particles from a fluid by driving the pulp to a **membrane** or **septum** (commonly called as the **filter medium**), porous to the fluid but impervious to the solid, through which the fluid called **filtrate** passes. The volume of filtrate collected per unit time is termed as the **rate of filtration**. As the filtration process proceeds, solid particles accumulate on the filter medium forming a packed bed of solids called **filter cake**. Pressure difference between two sides of filter medium, more pressure at the upstream side and less pressure at the downstream side, is the driving force.

The simplest type of filter consists of a tube of small bore through which the fluid is sucked while the solid particles accumulate at the entrance. As the device is operated, solids at first pass through the tube, but they quickly arch or bridge across the opening, allowing only clear liquid to pass afterwards as shown in Figure 19.3.

For bridging of a pore by solids, the diameter of the coarsest solid particles must exceed a certain minimum size; this minimum size is one-third of the pore opening if dealing with coarse material, but it may range to a small fraction of this ratio if the particles are fine. After bridging, solid particles continue to accumulate on the filter medium. The thickness of the cake thus increases as filtration continues and it offers more and more resistance to the flow of filtrate. The role of the filter medium is merely to act as a framework on which the bed of solids is supported. Although the primary purpose of the medium is to retain solids, other factors are significant. A good filter medium should be able to bridge solid particles across the pores, have a low resistance to filtrate flow, resist chemical attack, have sufficient strength to withstand the filtration pressure and mechanical wear and should allow efficient discharge of the cake.

Figure 19.3 Mechanism of filtration.

Rate of filtration depends on:

1 Filtering area.
2 Pressure difference across the cake and filter medium.
3 Average cross section of pores within the filter cake.
4 Number of pores per unit area of the filter medium.
5 Thickness of the filter cake.
6 Size range of the particles.
7 Degree of flocculation.
8 Pulp temperature.

19.2.1 Types of filters

Cake filters are most frequently used in mineral beneficiation plants where the recovery of large amounts of solids is the requirement. Cake filters are of two types: Pressure filters and Vacuum filters. Pressure filters are used where higher flow rates and better washing and drying are required. Vacuum filters are the most widely used filters in mineral beneficiation plants and they are of either drum type or disc type.

The **Rotary Vacuum Drum Filter** (Figure 19.4) consists of a cast cylinder mounted horizontally and rotated at a very low rpm. The cylinder has a perforated surface over which the filter medium is wrapped tightly. The periphery of the drum is divided into compartments. Each compartment is provided with a number of drain lines passing through the inside of the drum and connected to a rotary valve assembly on the central drum shaft. The drum is immersed to the required depth in the slurry which is kept agitated to prevent settling of solids. By the action of automatic rotary valve on the drum shaft, vacuum is applied to the immersed compartments and results in cake buildup on the filter medium surface. As the drum rotates, the cake is raised above the slurry level and wash water is sprayed on the surface if required. On further rotation,

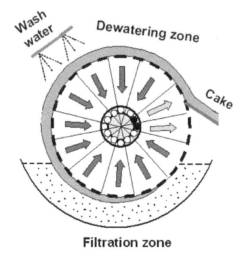

Figure 19.4 Rotary vacuum drum filter.

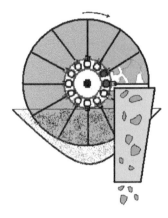

Figure 19.5 Rotary vacuum disc filter.   *Figure 19.6* Sector.

the cake is dried. The vacuum is continuously applied till the end of drying stage. Air is admitted in to the compartment before it re-enters in to the slurry, thus permitting the cake to be blown away from the filtering surface where it is removed by scraper on to a belt conveyor. The heart of the filter is the valve assembly which at a predetermined position causes a change from vacuum to pressure (or blowback).

The **Rotary Vacuum Disc Filter** (Figure 19.5) is similar to the drum filter except that instead of a cylindrical drum it consists of a number of circular discs mounted on a horizontal tubular shaft and spaced by means of hubs. A line of holes is drilled from the outside of the shaft into each of the tubes for the insertion of the sector nipple. Each disc consists of ten sectors (Figure 19.6). Filter medium consists of cloth bags that cover over the sector and fastened to the nipple. The completed sector is joined to the central shaft by screwing the nipple into the holes of the shaft.

The cycle of operation is similar to that of a drum filter. The build up of cake takes place on both sides of the disc in the disc filter whereas in the drum filter cake build up is on the periphery of the drum. Hence the filtering area in the disc filter is more for the same floor area. It can also filter several products in single unit. Worn out cloth can be replaced without interrupting the process for long time.

Ceramic Filters: Ceramic filter discs utilize microporous ceramic sectors instead of conventional filter cloth. Suction is created through capillary action. The microporous filter medium allows only liquid to flow through and no air penetrates the filter medium. Consequently, Ceramic filters require only a small vacuum pump to transfer filtrate from the discs to the filtrate receiver. Filter cake is removed from the ceramic discs by scraper, eliminating the need for compressed air for blow off.

The capillary action dewatering system is exceptionally versatile for many applications. The disc material is inert, resistant to almost all chemicals and slurry temperatures, and has a long operational life. Ceramic filters are best suited to filter feed slurries with consistent, high solids concentration, and to solids with a d80 coarser than 40 microns.

Chapter 20

Materials handling

The handling and storage of materials (Bulk solids) constitutes a major function in all Mineral Beneficiation plants. The handling of Bulk solids is required between each processing step in a plant, and often as part of a processing step. Surge capacity or storage is also required, both ahead of the plant and within it. As the costs of storage and handling systems in Beneficiation plants are substantial, the storage and handling facilities should be designed and operated with a view to obtaining maximum reliability, efficiency and economy. Bulk solids consist essentially of many particles or granules of different sizes (and possibly different chemical compositions and densities) randomly grouped together to form of a bulk. The nature of such a material – that is, its appearance, its feel, the way it behaves in various circumstances, and so on – is thus dependent upon many factors, but principally upon the size, shape and density of the constituent particles. The nature of bulk solids, described in terms of appropriate characteristics is an essential consideration when designing or selecting equipment for its handling or storage. Size, Shape and Surface area of the particles, Density of the particles, Bulk density, Compressibility, Cohesion and Adhesion, Angle of repose, Angle of fall, Angle of difference, Angle of spatula, Angle of surcharge, Angle of slide, Tackiness, Abrasion, Corrosion, Friability, Dispersability, Hygroscopicity and Moisture content, are some of the important properties of bulk solids to be considered while handling them. Storage and transport systems for bulk materials, slurries and water are inevitable for beneficiation plants.

20.1 PROPERTIES OF BULK SOLIDS

20.1.1 Size of the particles

Various terms like coarse, small, fine and superfine, are used to give a qualitative indication of the size of the particles constituting a bulk solid. The word 'size' is used loosely to mean some sort of average dimension across the particle.

There are many methods of determining the particle size distribution of bulk solids. The most popular (and cheapest) method of particle size analysis, especially with relatively coarse materials, is sieving. The sieve analysis method is described in chapter 4. Sedimentation, elutriation, Optical microscopy, Coulter counter and Laser diffraction spectrometry are some other methods.

20.1.2 Shape of the particles

Experience has shown that the shape of the constituent particles in a bulk solid is an important characteristic as it has a significant influence on their packing and flow behaviour. Flaky, needle shaped, bulky, rounded, angular, etc. are the terms used to indicate the shape of the particles qualitatively. Many attempts have been made to define the shape of non-spherical particles to indicate the extent to which particles differ from the sphere. Sphericity and the shape factor are the two terms used for the quantitative representation of the shape of the particles.

20.1.3 Surface area of the particles

The specific surface is defined as the sum of the surface areas of all particles contained in unit mass of the material. The surface area of certain finely divided bulk materials is of considerable importance during the processing and use of these materials. Various techniques have therefore been devised to measure this property. The most common type of instrument for measuring the surface area of powders and particulate materials is the permeameter.

20.1.4 Particle density (Density of the bulk solids)

The density of the particle is defined as the mass of the particle per unit volume. The ratio of the density of the particle to the density of water is defined as specific gravity. For a bulk material (bulk solids), the average particle density can be determined by dividing the mass of the material (solids) with the true volume occupied by the particles (not including the voids). The procedure for determination of the average particle density is described in chapter 11.

20.1.5 Bulk density

The bulk solids are really a combination of particles and space, the fraction of the total volume not occupied by the particles is referred as the 'voidage' or 'void fraction'. Sometimes the term 'porosity' is applied to bulk solids to mean the same as 'voidage'. The particle porosity can be defined as the ratio of the volume of pores within a particle to the volume of the particle (inclusive of pores).

A quantity of particulate or granular material will have an apparent density, usually termed as 'bulk density', which can be defined as the mass of the material divided by its total volume (particles and voids). Three kinds of bulk density that apply to materials handling calculations are (1) Aerated density (2) Packed density (3) Dynamic or Working density.

When the sample of the bulk material is carefully poured into a measuring cylinder to measure its volume, then the computed density is called as 'Aerated', 'loose', or 'poured' bulk density (ρ_a). If the sample is packed by dropping the cylinder vertically from a height of one or two centimeters on to a table, a number of times, then the computed density is called 'packed' or 'tapped' bulk density (ρ_c). The dynamic or working density (ρ_w) is a function of Aerated and Packed densities.

20.1.6 Compressibility

Compressibility (C) is a function of packed bulk density and aerated bulk density and expressed as:

$$C(\%) = \frac{100(\rho_c - \rho_a)}{\rho_c}$$

20.1.6.1

where
 C = compressibility
 ρ_c = packed bulk density
 ρ_a = aerated bulk density

The lower the percentage of compressibility, the material is more free flowing. The dividing line between free flowing (granular) and non-free flowing (powder) is about 20–21% compressibility. A higher percentage indicates a powder that is non-free flowing and will be likely to bridge in a hopper. The compressibility of a material often helps to indicate uniformity in the size and shape of the material, its deformability, surface area, cohesion and moisture content. The dynamic or working density (ρ_w) is expressed as:

$$\rho_w = (\rho_c - \rho_a) \, C + \rho_a$$

20.1.6.2

20.1.7 Cohesion and adhesion

Cohesion is defined as the molecular attraction by which particles of a body or material are united or held together. When the forces of attraction (cohesion) are low, the bulk material can be made to flow easily under the influence of gravity with the particles moving as individuals relative to one another. Dry sand is a familiar example of free-flowing bulk solids. However, high inter-particle cohesive forces may be caused by moisture or electrostatic charging in fine materials and tend to form agglomerates so that the material flows in an erratic manner as lumps.

Adhesion is the sticking together or adhering of substances in contact with each other. Cohesion is internal, adhesion is external. Adhesion describes the tendency of solid particles to 'stick' to a containing surface, such as a wall of a hopper or the side and bottom surfaces of a channel or chute. An extreme example of this is kaolin clay, which is so tacky that it will stick to a wall when thrown against it. This can create unusual problems in moving this material from storage. Adhesive materials tend to bridge in storage and thus require external assistance.

20.1.8 Angle of repose

Angle of repose is defined as the included angle formed between the edge of a cone shaped pile formed by dropping the material from a given elevation (6.8 cm – according to the standard) and the horizontal (Figure 20.1).

Angle of repose gives a direct indication of how free flowing the material will be. Angle of repose is sensitive to the condition of the supporting surface, the smoother the surface the smaller the angle. Moisture tends to increase the angle of repose. Bulk

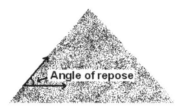

Figure 20.1 Angle of repose.

Table 20.1 Flowability character.

Angle of repose	Flow character
25–30°	Very free-flowing
30–38°	Free-flowing
38–45°	Fair flowing
45–55°	Cohesive
>55°	Very cohesive

solids with an Angle of Repose between 25° and 35° are generally considered free flowing. Table 20.1 shows flowability character based on angle of repose.

20.1.9 Angle of fall

When a material lies in a pile at rest, it has a specific Angle of Repose. If the supporting surface experiences vibrations, impacts or other movement, the material on the sloped sides of the pile will dislodge and flow down the slope. The new Angle of Repose that forms is referred to as the Angle of Fall. After measuring the angle of repose, the cone shaped pile of material is jarred by dropping a weight near it. The pile will fall resulting in a new, shallower angle with the horizontal. This new angle of repose is measured as angle of fall.

The way the pile falls is of special interest. If particles fall and spread out along the slope of the pile, only the degree of flowability is indicated. If the entire pile collapses, it indicates that the material contained entrained air and is prone to flushing.

20.1.10 Angle of difference

The angle of difference is the difference between the angle of repose and the angle of fall. The greater the angle of difference (between angles of repose and fall) the more free flowing the material is. It is an indirect measure of fluidity, surface area and cohesion.

20.1.11 Angle of spatula

The Angle of Spatula provides an indication of the internal friction between particles. It is determined by inserting a flat blade into a pile of granular material and lifting vertically. The new Angle of Repose, which the material forms relative to the blade surface, is known as the Angle of Spatula. Generally, bulk solids with an Angle of

Spatula less than approximately 40° are considered as free flowing. A highly flowable material will have an acute angle of spatula.

20.1.12 Angle of surcharge

The Angle of Surcharge is defined as the included angle formed between the edge of a cone shaped material and the horizontal when the material is at rest on a moving surface, such as a conveyor belt. This angle is usually 5° to 15° less than the angle of repose, though in some materials it may be as much as 20° less. The angle of surcharge is often called the dynamic angle of repose.

One yardstick to measure the flowability of a particular material is by the angle of repose and angle of surcharge. The inter-relationship is shown in Table 20.2.

20.1.13 Angle of slide

This is the angle of a flat surface on which a material will slide down due to its own weight. Angle of slide provides an indication of the material's flowability and is particularly useful in hopper and chute design.

20.1.14 Tackiness

Tackiness is the tendency of a material to stick to surfaces on which it is deposited and to cling to adjacent particles. A rough test for tackiness is to try to form a ball of the material by rolling it between the palms of the hands. If the ball falls apart, it is not tacky. If it sticks together, it is tacky.

Table 20.2 Inter-relationship between angle of repose, surcharge and flowability.

Very free flowing	Free flowing	Average flowing		Sluggish
5° angle of surcharge	10° angle of surcharge	20° angle of surcharge	25° angle of surcharge	30° angle of surcharge
0–20° angle of repose	20–30° angle of repose	30–35° angle of repose	35–40° angle of repose	40°–up angle of repose
Uniform size, very small rounded particle, either very wet or very dry, such as dry silica sand, cement	Rounded, dry polished particles, of medium weight	Irregular, granular or lumpy materials of medium weight, such as anthracite coal, clay, etc.	Typical common materials such as bituminous coal, stone, most ores, etc.	Irregular, stringy, fibrous, interlocking material, such as wood chips, tempered foundry sand, etc.

20.1.15   Abrasion

Abrasion means scraping off or wearing away. Knowing a material's abrasiveness is important in the proper design of equipment to protect against wear. The materials like coke and foundry sand will wear hoppers, chutes, screw feeders and conveyors, and pneumatic handling systems. Hardened steels, wear resistant liners and high density plastics must be considered for contact materials in such cases.

20.1.16   Corrosion

Corrosion can be defined as the deterioration of the material and its properties due to chemical or electrochemical reaction between a material, usually a metal, and its environment. Most metals corrode on contact with water (and moisture in the air), acids, bases, salts, oils, and other solid and liquid chemicals. When corrosive materials are processed, they must be handled in equipment with contact surfaces of alloy steel, special plastics or coated with corrosion resistant paint.

20.1.17   Friability (Degradation)

Friability means the breaking down of the particles in to smaller pieces. If it is undesirable to have any breakdown of the product in the process, as in coal, it is mandatory to use equipment whose design or performance will prevent such breakage.

20.1.18   Dispersibility

Dispersibility is the basic property that causes a material to flood or to produce dustiness in the surroundings. Dispersibility indicates the dusting and flushing characteristics of a material. It is a measure of the propensity for a granular material to form dust and thus lose mass to the surrounding air. Dispersible materials are generally of low bulk density and fine particle size, which causes them to behave more like a gas or a liquid than a solid. Dispersibility and floodability are interrelated. Materials with a dispersibility rating of more than 50% are very floodable and are likely to flush from a storage bin unless measures are taken to prevent this occurrence.

20.1.19   Moisture content and hygroscopicity

Materials that have an inherently high percentage of moisture may pose a handling problem. Free moisture, surface moisture or combined moisture as in wet ores cause the problems of sticking and poor flow. Generally, free moisture over 5–10% is considered risky. Particles of high porosity pick up and retain moisture and pose problems. Likewise, those materials that are hygroscopic or absorb moisture will cake and refuse to flow.

20.2   STORAGE

The following are the main purposes of ore storage:

1   To receive ore intermittently and deliver it smoothly.
2   To accumulate mill products for intermittent disposal.

3 To maintain an adequate tonnage of ore for treatment.
4 To smooth out irregularities in working (surge storage).
5 To facilitate balanced blending of dissimilar minerals.

Storage is accomplished in stock piles, bins, tanks and ponds.

Stockpiling is defined as the storage of dry bulk material in piles on the ground and is adopted to store coarse material of low value at outdoors, especially if the duration of storage is extensive. Stockpiles are formed on a concrete or earthen pad. Fixed, traveling and radial stackers, tripper and shuttle conveyors are used to form stockpiles. Cyclic machines such as grab bucket, dragline and shovel, continuous machines such as drum, rake and bucket wheel are used for reclaiming the stock piles. Stacker/re-claimers are also available to perform both functions.

Bins, Silos and **Bunkers** are used for dry or filtered materials where the storage period is short. A bin is a container of cylinder and hopper sections for bulk solids with one or more outlets for withdrawal either by gravity alone or by gravity assisted by flow promoting devices such as feeders. Silos, bunkers and other specialized terms used in various industries are also bins. A Silo has large cylindrical section whereas a Bunker has short cylindrical section.

Tanks, usually with an agitator, are employed for storing pulps and slurries.

Ponds and Dams are used for storing water and tailing slurries.

20.3 CONVEYING

The ore is transported by wagon, truck, rope haulage, aerial tramway etc., from mine to beneficiation plant. Within the mill, the ore is transported by gravity and by conveyors. **Gravity transport** is the flow of material in which actuating force is gravity and is carried out in chutes and launders. **Chutes** are steeply inclined troughs of rectangular sections for the gravity transport of dry solids. **Launders** are gently sloping troughs of rectangular, triangular or semicircular sections for the gravity transport of suspensions of ore or mineral.

Conveyors are used to transport the ore when the horizontal distance is relatively short. There is no arbitrary limit beyond which a conveyor system cannot be used. Travel may be horizontal or inclined either up or down, the maximum practicable inclination varies with the type of conveyor and the nature of materials.

A **Belt conveyor,** the most versatile and widely used type, consists of an endless belt running around head and tail pulleys and resting on various kinds of idlers at intervals along both its upper and return runs. Drive is generally through the head pulley.

Pan conveyors are similar to belt conveyors in method of drive and support, but differ in that the carrying surface consists of a series of articulated plates or shallow pans supported on rollers and tied together by pins.

A **Vibrating conveyor** consists of a material-transporting trough driven by a vibrator.

A **Flight conveyor** is essentially a trough through which a series of scrapers attached to chain or rope is drawn.

An **En masse** conveyor conveys materials by causing them to flow in a compact and unbroken stream through a conduit.

A **Screw conveyor** consists of a spiral blade attached to a revolving shaft which pushes the material along the bottom of a semicircular trough.

A **Bucket conveyor** consists of an endless belt or chain with buckets attached, running upon two pulleys or sprocket wheels, one above and one below. The belt or chain travels vertically or at a steep angle.

Skip hoists are widely used to haul ore from underground mines and to elevate coarse bulk materials over a limited distance. They provide an intermittent flow of material.

Pneumatic and **Slurry** transport are the methods of transporting material using air and water as the transporting medium respectively. Material to be transported is usually in powdered form and is transported through a pipeline system by utilizing the kinetic energy of the fluid (air/water) to move the material. This kinetic energy to the fluid is imparted by a blower or vacuum pump in the case of pneumatic transport and by a reciprocating or centrifugal pump in the case of slurry transport.

20.4 FEEDERS

Feeding is essentially a conveying operation in which the distance traveled is short and in which close regulation of the rate of passage is required. The feeder is installed at the outlet of a bin, silo or bunker. The different types of feeders are chain feeder, apron feeder, pan feeder, belt feeder, roller feeder, rotary feeder, reciprocating-plate feeder, plunger feeder, revolving-disk feeder, and vibrating feeders.

20.5 DISPOSAL OF PRODUCTS

The concentrated product from a beneficiation plant is transported through wagons, trucks, conveyors, and pipelines for further processing. The methods for disposal of tailings include discharge of tailing into rivers and streams, dumping of coarse tailings on to land, re-using tailings for some purpose, filling of mined-out areas etc. The disposal of tailings is a major environmental problem. It leads to pollution which is of major concern to the public and government. Uncontrolled discharge of tailings is no longer tolerated. One has to follow environmental legislation laid down by the government before the tailings are disposed of.

Chapter 21

Beneficiation of minerals

The Mineral Beneficiation plant for separating metalliferous mineral from gangue is called a **concentrator** or a **mill** and the process is called concentration or milling.

Minerals are beneficiated in several ways. The operations employed for the beneficiation of minerals depends on:

Liberation size
Physical properties such as;
    Specific gravity,
    Surface property,
    Magnetic susceptibility,
    Electrical conductivity etc.
Required grade of the ore
Operating characteristics of the equipment used
Financial feasibility

Detailed laboratory investigations are to be carried out in order to assess the various properties of the minerals present in the ore before planning for its beneficiation.

For the beneficiation of Iron ores, processes such as sizing, washing, gravity concentration like heavy medium separation, jigging, spiraling and magnetic separation are in use, after comminution to appropriate sizes.

Jigging and heavy medium separation operations are used for the beneficiation of Manganese ores. Spiraling and tabling are the major beneficiation operations for chrome ores.

In the case of lead, zinc and copper ores, valuables liberate when the ore is grounded down to 200 mesh. Froth flotation is employed for beneficiation of these ores.

Beach sand needs no comminution as the minerals are present in them as free particles. Mainly magnetic and electrostatic separation operations are used for separation of Ilmenite, Rutile, Zircon, Monazite etc.

Coal, is beneficiated principally by various types of heavy medium separation operations, Jigging, and flotation.

It is to be noted that it is not practically possible to achieve 100% liberation i.e., all the valuables will not detach during comminution. Hence the concentrate contains small amounts of gangue, and the tailing contains small amounts of valuable. In some cases, a middling product, containing considerable amount of valuables, is obtained. In such cases, the middling product is indicative of more locked particles.

Hence this product is again comminuted for more liberation and re-beneficiated to obtain valuables.

To get 100% liberation, all the ore is to be comminuted to a size far less than the grain size. This leads to consumption of more energy and time. Comminution operation becomes more costly. If beneficiation is done after this kind of comminution, the cost of the whole process will be high and may exceed the cost of the product. Hence such beneficiation is not practiced.

The various operations of a Beneficiation plant can be presented well in the form of a flow sheet. A flow sheet is a short way of representing the process adopted in a plant, equipment used and routing of the material, starting from raw material to finished product. Flow sheets vary greatly in form according to the information sought to be conveyed. The simplest form is to write various operations in sequence as employed in the plant (Figure 21.1). The most commonly used form is the flow sheet represented in simple block diagrams as shown in Figure 21.2. The flow sheet can be drawn in the diagrammatic form wherein the equipment used is represented by standard drawings and the flow path is represented by the arrows. Figure 21.3 shows the diagrammatic form of the Iron Ore Beneficiation Plant. Condensed flow sheets are useful for the study and comparison of practices followed in different parts of the mills.

The flow sheet is devised to make apparent at a glance the principal operations employed in the treatment scheme, the extent of each operation, the products made and the points at which they are taken out. The accessory apparatus such as storage bins, pumps, pipelines, conveyors, feeders, weighing equipment, measuring instruments,

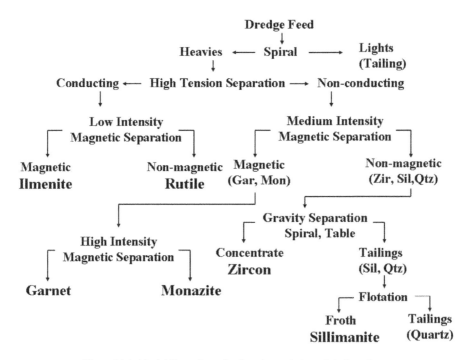

Figure 21.1 Model flow sheet for beach sands beneficiation plant.

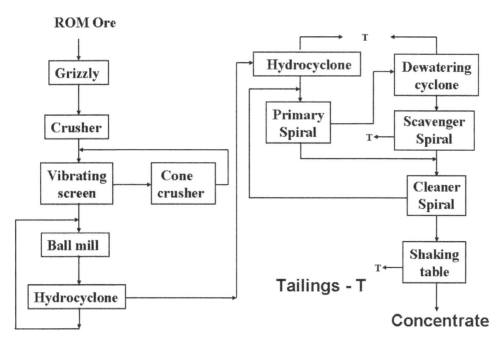

Figure 21.2 Model flow sheet for chromite ore beneficiation plant.

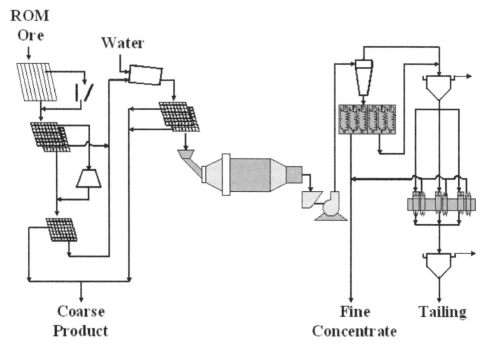

Figure 21.3 Model flow sheet for iron ore beneficiation plant.

sampling apparatus, etc., and the reagents used are sometimes denoted by numbers which indicate relative location and refer to details in the accompanying list.

Figure 21.1 shows a model flow sheet for Beach Sands Beneficiation Plant in the form of simple writing of various operations and materials.

Figure 21.2 shows a model flow sheet for a Chromite Ore Beneficiation plant in the form of block diagrams.

Figure 21.3 shows a model flow sheet for an Iron Ore Beneficiation Plant in the form of standard drawings.

21.1 AGGLOMERATION

Agglomeration is a method of bringing the fine particles of ore together and agglomerates them into lumps of suitable size and strength. Agglomeration is essential before feeding the ore into a metallurgical furnace as it requires ore at coarser sizes to have sufficient porosity of the ore bed in order to complete necessary reduction reactions for extraction of metal. Agglomeration is the ultimate method in preparing the feed for metallurgical furnaces. Agglomeration is very much essential in case of Iron Ores as about 50% of the ore mined is at fine size which needs to be agglomerated to utilize the fines for extraction. Hence agglomeration became a part of Mineral Preparation in addition to Beneficiation.

Briquetting, Sintering and Pelletization are the three important agglomeration methods. **Briquetting** consists of pressing of ore fines, with or without a binder, into a block or briquette of some suitable size and shape, and then subjecting it to a hardening process. Fines of coal, iron ore, chrome ore, etc. can be made into reasonably strong briquettes. **Sintering** is a process of heating a mass of fine particles to the stage of incipient fusion for agglomerating them into lumps. Iron ore fines are sintered before feeding to a Blast furnace. **Pelletization** consists of rolling of moist iron ore fines into balls, drying, preheating and firing to produce hardened balls which are an excellent feed for a blast furnace.

21.2 MATHEMATICAL MODEL

A quantitative description of a system or process by one or more mathematical equations is termed as **Mathematical Model** or simply **Model**. The Model is used to study the process behaviour with respect to the variables and parameters of interest using a digital computer. For a given system, the parameters are constants that characterize the system. If a system is described by a mathematical equation, a coefficient in that equation is a parameter. The variables are the attributes having arbitrary values. Some of the variables in a system are subject to the control and are referred as control variables.

The Mechanistic model, Population Balance model and Empirical or Statistical model are the three types of mathematical models. Empirical models are extensively used by Mineral Engineers to describe the processes. The Empirical model is derived by formulating the data obtained by experimentation. Models are useful to design, evaluate, optimize and control the processes.

21.3 SIMULATION

Simulation can be defined as a process of imitating a real phenomenon with a set of mathematical models (formulas). In theory, any phenomena that can be reduced to mathematical data and equations can be simulated on a computer. In practice, however, simulation is extremely difficult because most natural phenomena are subject to an almost infinite number of influences. One of the tricks to develop useful simulations, therefore, is to determine the most important factors.

In addition to imitating processes to see how they behave under different conditions, simulations are also used to test new theories. After creating a theory of causal relationships, it is codified in the form of a computer program. If the program then behaves in the same way as the real process, there is a good chance that the proposed relationships are correct.

21.4 AUTOMATIC CONTROL

Automatic control in a Mineral Beneficiation plant is aimed at one or more of the following:

1   Stabilization and optimization of the process.
2   Increased throughput.
3   Improved recovery of the valuable minerals.
4   Improved concentrate grade.
5   Reduced operating costs.

Automatic control achieves the regulation of the process with measuring and recording instruments such as Belt scales to determine the mass transported by a belt conveyor, Magnetic or ultrasonic flow meters to measure the flow rate of slurries, Nuclear density meters to determine pulp density, Optical or ultrasonic analyzers to determine particle size distribution, X-ray fluorescence analyzer for on-stream assaying etc. Other important sensors are pH meters, and level and pressure transducers. Most control systems installed today are based on a digital computer. It is a high speed machine which uses numbers rather than physical quantities when processing data.

Encyclopedia of Immunology

Chapter 22

Applications

Various formulae indicated in the text and their applications are illustrated by considering suitable problems. Problems are solved from the basic definitions of various quantities in order to have clear understanding. Wherever the formulae are used, they are derived from the fundamentals.

22.1 ASSAY VALUE AND GRADE

The assay value of an ore, and the products (concentrate, middling and tailing), is determined by chemical analysis or by other instrumental methods after collecting samples from them. Calculation of the percent of metal present in a mineral is shown in chapter 1.2. The percent metal present in a few minerals is calculated and indicated in Table 1.2. These are the maximum metal values in respective minerals. The grade of any concentrate obtained from beneficiation operation cannot have more than these values. In the case of industrial minerals the percentage of the specific component required is calculated. Percent CaO in Limestone, percent Al_2O_3 and SiO_2 in china clay are the examples.

   The grade of a metallic ore is usually indicated by percent metal as 60% Fe Iron ore. It does not mean that 40% of ore is gangue minerals. This aspect is shown in Illustration 1.

Illustration 1: *An Iron Ore of 60% Fe contains only one Iron Mineral Hematite. What is the percent gangue in Iron Ore?*

Solution:

Chemical Formula of Hematite		$= Fe_2O_3$
Atomic weight of Iron		$= 55.85$
Atomic weight of Oxygen	$= 16.00$	
Molecular weight of Hematite	$= 55.85 \times 2 + 16.00 \times 3$	
		$= 159.70$
Percent Iron	$= \dfrac{55.85 \times 2}{159.70} \times 100$	$= 69.94$
% Fe in Hematite Fe_2O_3		$= 69.94\%$
% Fe in Iron Ore		$= 60\%$
$\therefore$ % Fe_2O_3 in Iron Ore		$= \dfrac{60.00}{69.94} \times 100$
		$= 85.79\%$
% gangue in Iron Ore		$= 100 - 85.79$
		$= 14.21\%$

22.2   SIEVE ANALYSIS

The method of sieve analysis, determination of average size of the sample and 80% passing size are indicated in chapters 4.1, 4.2 and 4.3. Determination of average size of the sample and 80% passing size are shown in Illustration 2.

Illustration 2: *From the sieve analysis data of a sample in Table 22.1, calculate (a) Average size of the sample and (b) 80% passing size.*

Table 22.1 Sieve analysis data.

Mesh number	Mesh size microns	Direct wt % retained
+18	853	7.0
−18 + 25	599	10.4
−25 + 36	422	14.2
−36 + 52	295	13.6
−52 + 72	211	9.2
−72 + 100	152	8.1
−100 + 150	104	8.2
−150 + 200	74	5.1
−200		24.2

Solution:

Average sizes for each fraction (d_i) and (w_i/d_i) are calculated and tabulated in Table 22.2:

Table 22.2 Calculated values for determination of average size.

Size of each fraction d_i	wt % w_i	w_i/d_i
853.0	7.0	0.0082
$(853 + 599)/2 = 726.0$	10.4	0.0143
$(599 + 422)/2 = 510.5$	14.2	0.0278
$(422 + 295)/2 = 358.5$	13.6	0.0379
$(295 + 211)/2 = 253.0$	9.2	0.0363
$(211 + 152)/2 = 181.5$	8.1	0.0446
$(152 + 104)/2 = 128.0$	8.2	0.0641
$(104 + 74)/2 = 89.0$	5.1	0.0573
$(74 + 0)/2 = 37.0$	24.2	0.6541
		0.9446

$\therefore$ Average size of the sample $= \dfrac{100}{\sum \frac{w_i}{d_i}}$ 4.2.1

$$= \frac{100}{0.9446} = 105 \ \mu m$$

For determination of 80% passing size, cumulative weight percentages passing are to be calculated by adding the weight percentages from the bottom. All the values are shown in the Table 22.3.

Table 22.3 Calculated values for drawing graph.

Mesh number	Mesh size microns	wt % retained	Cumulative wt % passing
+18	853	7.0	100.0
−18	853	10.4	93.0
−25	599	14.2	82.6
−36	422	13.6	68.4
−52	295	9.2	54.8
−72	211	8.1	45.6
−100	152	8.2	37.5
−150	104	5.1	29.3
−200	74	24.2	24.2

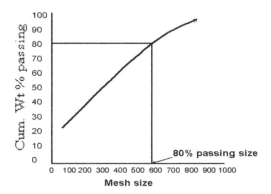

Figure 22.1 Graph to determine 80% passing size.

To determine 80% passing size, a graph between cumulative wt% and mesh size is drawn as shown in Figure 22.1.

A horizontal line is drawn at 80% cumulative weight percent passing. From the point of intersection of this horizontal line with the curve, a vertical line is drawn on to the x-axis. The point at which the vertical line meets x-axis is the 80% passing size. It is 580 microns.

22.3 DISTRIBUTION OF METAL VALUES

Metal values present in the ore vary with the size of the particle which depends on the liberation characteristics of the ore. To determine the metal values in different sizes of the ore particles, the sample of ore is to be subjected to the sieve analysis and each size fraction is to be analyzed for its metal value. From these values, percent metal present in the ore can be calculated. The details are shown in Illustration 3.

Illustration 3: *Iron ore fines from Bailadila have been grounded in a Ball mill to −1000 micron size. A sample is collected from the Ball mill product and analysed for its size distribution and metal values. The data is given in Table 22.4:*

Table 22.4 Size analysis data.

Size, microns	Mill product Wt %	% Fe Wt %
−1000 + 850	02.97	67.01
−850 + 500	01.53	65.22
−500 + 212	07.88	63.63
−212 + 150	04.60	63.37
−150 + 075	29.86	63.98
−075 + 045	12.03	66.05
−045	41.13	64.66

Calculate the %Fe in Ball mill product.

Solution:

The third column of Table 22.5 was converted into fraction of Fe. The fourth column of Table 22.5 is the simple multiplication of column 2 and column 3. The total of this column gives % Fe in the Ball Mill product.

Table 22.5 Calculated values.

Size, microns	Mill product Wt %	Fraction of Fe Weight fraction	Fe content Weight
1	2	3	4
−1000 + 850	02.97	0.6701	1.990
−850 + 500	01.53	0.6522	0.998
−500 + 212	07.88	0.6363	5.014
−212 + 150	04.60	0.6337	2.915
−150 + 075	29.86	0.6398	19.104
−075 + 045	12.03	0.6605	7.946
−045	41.13	0.6466	26.594
	100.00		64.561

$\therefore$ %Fe in the Ball Mill product = 64.561%

22.4 EFFICIENCY OF THE SCREEN

A screen is said to behave perfectly if, in a mixture of different sizes of materials, all material of a particular size less than the screen aperture is separated from the mixture. In general, absolute separation of different sized particles using a screen is difficult as it involves probabilities of movement of particles at different stages that may be difficult to determine. Hence it is necessary to express the efficiency of the process.

Screen efficiency (often called the effectiveness of a screen) is a measure of the success of a screen in closely separating oversize and undersize materials. There is no standard method for defining the screen efficiency. Depending on whether one is interested in removing oversize or undersize material, screening efficiencies may be defined in a number of ways.

Screen efficiency can be calculated based on the amount of material recovered at a given size. In an industrial screening operation, it is to be specified whether the required material is oversize or undersize or both. For the oversize material, the screen efficiency (η_c) is defined as the ratio of weight of actual oversize material present in the feed to the weight of overflow material obtained from the screen. For the undersize material, the screen efficiency (η_u) is defined as the ratio of weight of underflow material obtained from the screen to the weight of actual undersize material present in the feed.

In an industrial screen, if there are no broken or deformed apertures and screen is perfectly made, no single coarse particle coarser than the size of aperture pass through. Therefore aforesaid definitions are applicable under such assumed conditions. In reality, some coarse particles, which may be less in quantity, will report to underflow fraction. Under such cases equations for efficiency can be derived by writing mass balance equations for the screening operation as follows:

Mass balance of total material $F = C + U$ 22.4.1

Mass balance of oversize material $Ff = Cc + Uu$ 22.4.2

Mass balance of undersize material $F(1 - f) = C(1 - c) + U(1 - u)$ 22.4.3

where

    F = Weight of the total material in the feed
    C = Weight of the overflow material obtained from the screen
    U = Weight of the underflow material obtained from the screen
    f = fraction of oversize material in the feed
    c = fraction of oversize material in the overflow obtained from the screen
    u = fraction of oversize material in the underflow obtained from the screen

On computation of equations 22.4.1 and 22.4.2 $\dfrac{C}{F} = \dfrac{f - u}{c - u}$ 22.4.4

$$\text{and } \frac{U}{F} = \frac{c - f}{c - u} \qquad 22.4.5$$

The recovery of oversize material into the screen overflow is referred as Screen Efficiency (or Screen Effectiveness), η_c, based on the oversize material

$$\eta_c = \frac{Cc}{Ff} = \frac{c(f - u)}{f(c - u)} \qquad 22.4.6$$

The recovery of undersize material into the screen underflow is referred to as Screen Efficiency (or Screen Effectiveness), η_u, based on the undersize material:

$$\eta_u = \frac{U(1 - u)}{F(1 - f)} = \frac{(1 - u)(c - f)}{(1 - f)(c - u)} \qquad 22.4.7$$

A combined overall efficiency, or overall effectiveness, η, is then obtained by multiplying equations 22.3.6 and 22.3.7:

$$\eta = \eta_c \times \eta_u = \frac{c(f-u)(1-u)(c-f)}{f(c-u)^2(1-f)} \qquad \text{5.6.1}$$

If f, c, and u are expressed in terms of the fractions of undersize material in feed, overflow and underflow respectively, the following are the formulae:

Mass balance of total material $F = C + U$ 22.4.8

Mass balance of undersize material $Ff = Cc + Uu$ 22.4.9

Mass balance of oversize material $F(1-f) = C(1-c) + U(1-u)$ 22.4.10

On computation of equations 22.4.8 and 22.4.9 $\dfrac{C}{F} = \dfrac{f-u}{c-u}$ 22.4.11

$$\text{and } \frac{U}{F} = \frac{c-f}{c-u} \qquad \text{22.4.12}$$

The recovery of oversize material into the screen overflow is referred as Screen Efficiency (or Screen Effectiveness), η_c, based on the oversize material:

$$\eta_c = \frac{C(1-c)}{F(1-f)} = \frac{(1-c)(f-u)}{(1-f)(c-u)} \qquad \text{22.4.13}$$

The recovery of undersize material into the screen underflow is referred as Screen Efficiency (or Screen Effectiveness), η_u, based on the undersize material

$$\eta_u = \frac{Uu}{Ff} = \frac{u(c-f)}{f(c-u)} \qquad \text{22.4.14}$$

A combined overall efficiency, or overall effectiveness, η, is then obtained by multiplying the equations 22.4.13 and 22.4.14:

$$\eta = \eta_c \times \eta_u = \frac{u(u-f)(1-c)(f-c)}{f(u-c)^2(1-f)} \qquad \text{5.6.2}$$

Efficiency of the screen is calculated by using both the formulae 5.6.1 and 5.6.2 in Illustration 4.

Illustration 4: *A quartz mixture is screened through a 1.5 mm screen to obtain +1.5 mm fraction. The size analysis of feed, overflow and underflow is shown in Table 22.6:*

Table 22.6 Size analysis data.

Screen size mm	Weight percent retained this size		
	Feed	Overflow	Underflow
3.3	3.5	7.0	–
2.3	13.5	36.0	–
1.5	33.0	37.0	15.0
1.0	22.7	13.0	43.0
0.8	16.0	4.0	25.0
0.6	5.4	3.0	8.0
0.4	2.1	–	3.0
0.2	1.8	–	2.0
–0.2	2.0	–	4.0

Calculate the effectiveness of the screen.

Solution:

Fraction of +1.5 mm material in the feed $= f$

$= 3.5 + 13.5 + 33.0$ $= 50\% \Rightarrow 0.50$

Fraction of +1.5 mm material in the overflow product $= c$

$= 7.0 + 36.0 + 37.0$ $= 80\% \Rightarrow 0.80$

Fraction of +1.5 mm material in the underflow product $= u$

$= 15\% \Rightarrow 0.15$

Efficiency (effectiveness) of the screen $= \eta = \dfrac{c(f-u)(1-u)(c-f)}{f(c-u)^2(1-f)}$ 5.6.1

$= \dfrac{0.80(0.50-0.15)(1.00-0.15)(0.80-0.50)}{0.50(0.80-0.15)^2(1.00-0.50)} = 0.676 \Rightarrow 67.6\%$

Alternatively

Fraction of –1.5 mm material in the feed $= f$

$= 22.7 + 16.0 + 5.4 + 2.1 + 1.8 + 2.0$ $= 50\% \Rightarrow 0.50$

Fraction of –1.5 mm material in the overflow product $= c$

$= 13.0 + 4.0 + 3.0$ $= 20\% \Rightarrow 0.2$

Fraction of –1.5 mm material in the underflow product $= u$

$= 43.0 + 25.0 + 8.0 + 3.0 + 2.0 + 4.0$ $= 85\% \Rightarrow 0.85$

Efficiency (effectiveness) of the screen $= \eta = \dfrac{u(u-f)(1-c)(f-c)}{f(u-c)^2(1-f)}$ 5.6.2

$= \dfrac{0.85(0.85-0.50)(1-0.20)(0.50-0.20)}{0.50(0.85-0.20)^2(1-0.5)} = 0.676 \Rightarrow 67.6\%$

22.5 LAWS OF COMMINUTION

Bond's law of comminution is discussed in chapter 7.2. Illustration 5 is an example of how the power requirement for a comminution operation can be estimated.

Illustration 6 is an example to determine the work index if power consumption is known.

Illustration 5: *What is the power required to crush 100 tons/hr of limestone if 80% of the feed passes 2″ screen and 80% of the product passes 1/8″ screen? The work Index for limestone may be taken as 12.74 kWhr/short ton.*

Solution:
80% passing size of the product = P = 1/8″ = 3175 microns;
80% passing size of the feed = F = 2″ = 50800 microns;
Work Index of the limestone = W_i = 12.74 kWhr/short ton

As per the Bond's Law $W = 10\ W_i \left(\dfrac{1}{\sqrt{P}} - \dfrac{1}{\sqrt{F}} \right)$ 7.2.2

$$= 10 \times 12.74 \left(\frac{1}{\sqrt{3175}} - \frac{1}{\sqrt{50800}} \right)$$
$$= 1.696 \text{ kWhr/short ton}$$

Power per 100 tons/hr = 1.696 × 100 × 1.1023 = 186.95 kW
[∵ 1 metric ton = 1.023 × short ton]

Illustration 6: *If a ball mill requires 3 kWhr per ton of feed to reduce a feed of 1600 micron size to a product of 400 microns size, calculate the work index.*

Solution:
80% passing size of the product = P = 400 microns;
80% passing size of the feed = F = 1600 microns;
Power required = W = 3 kWhr/ton = 3/1.023 = 2.7 kWhr/ short ton

As per the Bond's Law $W = 10\ W_i \left(\dfrac{1}{\sqrt{P}} - \dfrac{1}{\sqrt{F}} \right)$ 7.2.2

$$2.7 = 10\ W_i \left(\frac{1}{\sqrt{400}} - \frac{1}{\sqrt{1600}} \right)$$
$$\Rightarrow W_i = 10.8 \text{ kWhr per short ton}$$

22.6   REDUCTION RATIO

Reduction ratio of any crusher can be determined from the sieve analysis of the feed and product of that crusher. Procedure for the determination of the average size of the sample is shown in Illustration 2. After determining the average size of the feed and product, the reduction ratio can be determined by using the following definition:

$$\text{Average Reduction Ratio} = \frac{\text{Average size of the feed particles}}{\text{Average size of the product particles}}$$

22.7 SIZE OF ROLL IN ROLL CRUSHER

By using the equations 8.3.1 and 8.3.2, the size of the roll can be determined as shown in Illustration 7.

Illustration 7: *If the coefficient of friction between rock and steel is 0.4, what is the minimum diameter of the roll to reduce 1.5" pieces of rock to 0.5".*

Solution:

Coefficient of friction $= \mu = \tan n/2 = 0.4$ 8.3.2

$\Rightarrow n/2 = 21°48'$

$\therefore \cos n/2 \qquad\qquad\qquad = 0.9285$

But $\cos \dfrac{n}{2} \qquad\qquad = \dfrac{D+s}{D+d}$ 8.3.1

$\Rightarrow 0.9285 \qquad\qquad = \dfrac{D+0.5}{D+1.5}$

$\Rightarrow D \qquad\qquad\qquad = 12.5"$

Minimum diameter of the roll required $= 12.5"$

22.8 CRUSHING CIRCUITS

As discussed in chapter 8.8, crushing may be performed either in an open circuit or in a closed circuit. Usually crushing is performed in a closed circuit with a screen. The screen separates the oversize material from the crusher product and it is re-fed to the same crusher. The overflow material obtained from the screen, which is re-fed to the crusher, is called circulating load. The method of calculation of the circulating load is shown in Illustration 8.

Illustration 8: *The crusher with 100 tons per hour of feed is operated in closed circuit with a screen of 2 inch opening. The crusher product contains 46% of +2 inch material. Draw the flow diagram and find the circulating load assuming the screen efficiency of 85% based on oversize material. Also find the circulating load assuming the screen efficiency of 85% based on undersize material.*

Solution:

The flow diagram is shown in Figure 22.2.
If screen efficiency is based on oversize material:

Let C be the circulating load.
Load on the gyratory crusher $= C + 100$ tons/hr
+2" material in crusher product $= 0.46 (C + 100)$ tons/hr

$$\text{Screen efficiency} = \eta = 0.85 = \frac{+2\text{" material present in the feed to the screen}}{\text{Overflow material obtained from the screen}}$$

$$= \frac{0.46(C+100)}{C}$$

$$\Rightarrow C = 117.95 \text{ tons/hr}$$

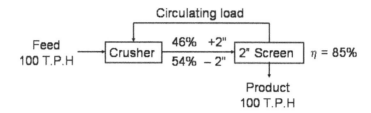

Figure 22.2 Closed circuit crushing.

If screen efficiency is based on undersize material:

Let C be the circulating load.
Load on the gyratory crusher = $C + 100$ tons/hr
–2″ material in crusher product = 0.54 $(C + 100)$ tons/hr

$$\text{Screen efficiency} = \eta = 0.85 = \frac{\text{Underflow material obtained from the screen}}{-2'' \text{ material present in the feed to the screen}}$$

$$= \frac{100}{0.54(C + 100)}$$

$$\Rightarrow C = 117.87 \text{ tons/hr}$$

22.9   CRITICAL SPEED

The speed at which the centrifuging of balls takes place in a ball mill is known as the **critical speed.**
  Let

m = mass of the ball, kg
R = Radius of the mill shell, metre
N = speed of the mill, rpm
v = linear velocity of the ball, m/sec = $\dfrac{2\pi RN}{60}$

  Let us consider a ball which is lifted up the shell. When the ball is moving around inside the mill, it follows a circular path till it reaches a certain height and then it changes its circular path and follows a parabolic path while dropping on to the toe (Figure 22.3).
  When the ball is at the point of changing its path, centrifugal and centripetal forces acting on it will just balance (Figure 22.4). The centrifugal force acting on the ball is $\frac{mv^2}{R}$. The component of weight of the body opposite to the direction of centrifugal force is the centripetal force i.e. mg cos α, where α is the angle made by the line of action of centripetal force with the vertical. P is a point where the ball

changes its circular path to parabolic path. At this point, both the forces are equal. Therefore,

$$\frac{mv^2}{R} = mg\cos\alpha \qquad\qquad 22.9.1$$

$$\Rightarrow \frac{\left(\frac{2\pi RN}{60}\right)^2}{R} = g\cos\alpha$$

$$\Rightarrow N^2 = \frac{(60)^2 g\cos\alpha}{4\pi^2 R} \qquad\qquad 22.9.2$$

If D is the diameter of the mill, d is the diameter of the ball, the radius of the circular path of outer most ball is $\frac{(D-d)}{2}$. On substitution:

$$N^2 = \frac{(60)^2 g\cos\alpha}{2\pi^2(D-d)} \qquad\qquad 22.9.3$$

Critical speed of the mill occurs when $\alpha = 0$. i.e. the ball will rotate along with the mill. At this point $\cos\alpha = 1$. Then $N = N_c$.

$$\therefore N_c^2 = \frac{(60)^2 g}{2\pi^2(D-d)} \qquad\qquad 22.9.4$$

If D and d are taken in metres, g is 9.81 m/sec², equation 22.9.4 is reduced to:

$$\text{Critical speed} = N_c = \frac{42.3}{\sqrt{(D-d)}} \text{ revolutions/minute} \qquad\qquad 9.1.1$$

Illustration 9: *What is the critical speed of a ball mill of 48″ I.D. charged with 3″ balls?*

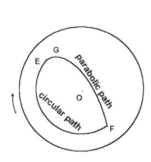

Figure 22.3 Path of a ball.

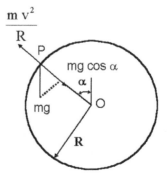

Figure 22.4 Forces on a ball.

Solution:
Diameter of the ball mill = 48″ = 48 × 0.0254 = 1.2192 metres
Diameter of the balls = 3″ = 3 × 0.0254 = 0.0762 metre

$$\text{Critical speed of the ball mill} = N_c = \frac{42.3}{\sqrt{(D-d)}} \qquad 9.1.1$$

$$= \frac{42.3}{\sqrt{(1.2192-0.0762)}} = 39.57 \text{ rpm}$$

Illustration 10: *Calculate the operating speed of a ball mill of 2 metres diameter containing steel balls (specific gravity of 7.6) of 0.1 metre diameter. The mill operates at 70% of the critical speed.*

Solution:
Diameter of the ball mill = 2 m
Diameter of the balls = 0.1 m

$$\text{Critical speed of the ball mill} = N_c = \frac{42.3}{\sqrt{(D-d)}} \qquad 9.1.1$$

$$= \frac{42.3}{\sqrt{(2-0.1)}} = 30.69 \text{ rpm}$$

Operating speed = 70% of critical speed = 0.7 × 30.69 = 21.48 rpm

22.10   GRINDING CIRCUITS

As discussed in chapter 9.6, grinding may be performed either in an open circuit or in a closed circuit. Determination of percent circulating load in closed circuit grinding operation is shown in Illustration 11.

Illustration 11: *A ball mill, receiving 100 dry tons of new crude ore per hour, is in operation in closed circuit with a classifier. The percent solids by weight in the feed to the classifier, in the classifier overflow (fines) and sands (coarse) are 50, 25 and 84 respectively. Calculate the percent circulating load.*

Solution:
The closed circuit grinding operation is as shown in Figure 22.5.

Fraction of solids in feed $= f = 50\% = 0.50$
Fraction of solids in overflow $= p = 25\% = 0.25$
Fraction of solids in underflow $= u = 84\% = 0.84$

In a closed circuit operation, New feed = Product = 100 dry tons/hr

Total weight of overflow (water + solids) pulp $= P = \dfrac{100}{0.25} = 400$ tons/hr

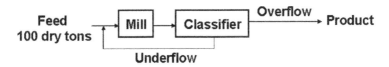

Figure 22.5 Closed circuit grinding.

Let

F = weight of feed pulp to the classifier

U = weight of underflow pulp from the classifier

Total pulp (solids + water) balance equation is $F = P + U \Rightarrow F = 400 + U$

Dry solids balance equation is $F f = P p + U u \Rightarrow F (0.5) = 400 (0.25) + U (0.84)$

Solving above two equations $U = 294.12$ tons/hr

Circulating load = Solids in underflow = $0.84 \times 294.12 = 247.1$ dry tons/hr

$$\% \text{ circulating load} = \frac{\text{Circulating load}}{\text{New feed}} \times 100 = \frac{247.1}{100} \times 100 = 247.1\%$$

22.11   DENSITY AND % SOLIDS

Density and percent solids calculations are illustrated by considering different problems in Illustrations 12, 13, 14 and 15.

Illustration 12: *When a slurry sample of 2 litres with solids specific gravity of 4.8 is filtered, 250 grams of filter cake is obtained. If the dry weight of the solids is 180 grams, Calculate*

a   *% solids by volume in the slurry*

b   *%solids by weight in the slurry*

c   *Liquid solid ratio by volume & by weight*

d   *Density of the slurry*

e   *% moisture in the filter cake*

f   *Bulk density of the filter cake*

Solution:

Volume of solids	$= 180/4.8 = 37.5$ cm³
Volume of water in the slurry	$= 2000 - 37.5 = 1962.5$ cm³
% solids by volume	$= \dfrac{37.5}{2000} \times 100 = 1.875\%$
Weight of water	$= 1962.5$ gm
Weight of solids	$= 180$ gm
Weight of the slurry	$= 1962.5 + 180 = 2142.5$ gm
% solids by weight	$= \dfrac{180}{2142.5} \times 100 = 8.4\%$
Liquid solid ratio by volume	$= \dfrac{1962.5}{37.5} = 52.3$

Liquid solid ratio by weight	$= \dfrac{1962.5}{180} = 10.9$
Slurry density	$= \dfrac{2142.5}{2000} = 1.07125$ gm/cm³
Weight of moisture in filter cake	$= 250 - 180 = 70$ gm
% moisture by weight in filter cake	$= \dfrac{70}{250} \times 100 = 28\%$
Volume of filter cake	$= 37.5 + 70 = 107.5$ cm³
Bulk density	$= \dfrac{250}{107.5} = 2.33$ gm/cm³

Illustration 13: *It is required to prepare a suspension of specific gravity 1.45 by adding sand of specific gravity 2.6 to the water. How much percent of sand by weight is necessary to prepare the suspension.*

Solution:
Equation 11.2 can be used to determine % sand:

$$\frac{C_w}{\rho_p} + \frac{1 - C_w}{\rho_w} = \frac{1}{\rho_{sl}} \qquad\qquad 11.2$$

$$\Rightarrow \frac{C_w}{2.6} + \frac{1 - C_w}{1.0} = \frac{1}{1.45}$$

$$\Rightarrow \% \text{ sand} = C_w = 0.5043 = 50.43\%$$

Illustration 14: *In a flotation plant, pulp of 40% solids by weight is conditioned for 5 minutes before pumped to flotation cells. If the solids are treated at the rate of 500 tons/hr and specific gravity of solids is 2.7, calculate the volume of conditioning tank required.*

Solution:

$$\text{Volumetric flow rate of solids} = \frac{500 \times 1000}{2700} = 185.2 \text{ m}^3/\text{hr}$$

$$\text{Dilution ratio} = \frac{1 - C_w}{C_w} \qquad\qquad 11.6$$

$$= \frac{1 - 0.40}{0.40} = 1.5$$

Mass flow rate of water	$=$ Mass flowrate of solids × dilution ratio
	$= 500 \times 1.5 = 750$ tons/hr
Volumetric flow rate of water	$= \dfrac{750 \times 1000}{1000} = 750$ m³/hr
Volumetric flow rate of slurry	$= 750 + 185.2 = 935.2$ m³/hr
Retention time	$= 5$ minutes

Volume of conditioning tank $= 935.2 \times \dfrac{5}{60} = 77.9 \text{ m}^3$

Illustration 15: *Two liquids of specific gravities 1.26 and 1.6 are to be mixed to obtain 300 cc solution of specific gravity 1.4. Calculate the quantities of two liquids required by volume and by weight.*

Solution:
Density of liquid 1 $= \rho_1 = 1.26$
Density of liquid 2 $= \rho_2 = 1.60$
Density of solution $= \rho_{12} = 1.40$

Let the fraction of liquid 1 by volume $= C_v$

$C_v \rho_1 + (1 - C_v) \rho_2 = \rho_{12}$ 11.7
$\Rightarrow C_v \times 1.26 + (1 - C_v) \times 1.6 = 1.4$
$\Rightarrow C_v = 0.588$

Fraction of liquid 2 by volume $= 1 - 0.588 = 0.412$

Volume of liquid 1 $= 0.588 \times 300 = 176.4$ cc
Volume of liquid 2 $= 0.412 \times 300 = 123.6$ cc
Let the fraction of liquid 1 by weight $= C_w$

$\dfrac{C_w}{\rho_1} + \dfrac{1 - C_w}{\rho_2} = \dfrac{1}{\rho_{12}}$ 11.2

$\Rightarrow \dfrac{C_w}{1.26} + \dfrac{1 - C_w}{1.6} = \dfrac{1}{1.4}$
$\Rightarrow C_w = 0.5294$

Fraction of liquid 2 by weight $= 1 - 0.5294 = 0.4706$

Total weight of the liquid $= 1.4 \times 300 = 420$ gm

Weight of liquid 1 $= 0.5294 \times 420 = 222.3$ gm
Weight of liquid 2 $= 0.4706 \times 420 = 197.7$ gm

22.12 SETTLING VELOCITIES

As discussed in chapter 12.1, settling velocity, size ranges of the particles in classifying products, and the velocity of hydraulic water employed in classifier are calculated in Illustrations 16, 17 and 18.

Illustration 16: *Calculate the terminal settling velocity of galena particle of 7.5 specific gravity and 10 microns in size settling in water.*

Solution:
Density of the particle $= \rho_p = 7.5$ gm/cc
Diameter of the particle $= d = 10$ microns $= 0.001$ cm
Density of water $= \rho_w$ $= \rho_f = 1.0$ gm/cc

Viscosity of water $= \mu_w$ $= \mu_f = 0.01$ poise

As per Stokes law $v_m = \dfrac{d^2 g(\rho_p - \rho_f)}{18\mu_f}$

$\qquad = \dfrac{(0.001)^2 (980)(7.5 - 1.0)}{18(0.01)} = 0.0354$ cm/sec

12.1.5

Illustration 17: *Quartz and galena particles at size range of 5.2 to 25 microns are present in a mixture. Determine the size ranges of pure quartz, pure galena and the third product of mixture which can be obtained if separated in a free settling classifier. The specific gravities of quartz and galena are 2.6 and 7.5 respectively.*

Solution:
Density of water $= \rho_w = 1.0$ gm/cc
Density of quartz $= \rho_q = 2.6$ gm/cc
Density of galena $= \rho_g = 7.5$ gm/cc

Let d_q and d_g be the sizes of quartz and galena particles.

The size of the galena particle that settles equally with the largest quartz particle:

$$= d_g = \left(\dfrac{\rho_q - \rho_w}{\rho_g - \rho_w} \right)^{1/2} d_q$$

12.1.9

$$\Rightarrow \quad d_g = \left(\dfrac{2.6 - 1.0}{7.5 - 1.0} \right)^{1/2} \times 25 = 12.4 \ \mu m$$

The size of the quartz particle that settles equally with the smallest galena particle:

$$= d_q = \left(\dfrac{\rho_g - \rho_w}{\rho_q - \rho_w} \right)^{1/2} d_g$$

12.1.9

$$\Rightarrow d_q = \left(\dfrac{7.5 - 1.0}{2.6 - 1.0} \right)^{1/2} \times 5.2 = 10.4 \ \mu m$$

Therefore:

1   All galena particles of size more than 12.4 microns will be obtained as pure galena fraction from the free settling classifier.
2   All quartz particles of size less than 10.4 microns will be obtained as pure quartz fraction from the free settling classifier.
3   The third fraction i.e. the mixture contains:

Quartz particles of size 10.4 – 25 microns
Galena particles of size 5.2 – 12.4 microns

Illustration 18: *A mixture having size range of 40 to 90 microns contains two minerals of specific gravities 7.0 and 2.0. Can you separate the two minerals completely by using free settling hydraulic classifier? If so explain how.*

Solution:

Density of water $= \rho_w = 1.0$ gm/cc
Density of heavy mineral $= \rho_h = 7.0$ gm/cc
Density of light mineral $= \rho_l = 2.0$ gm/cc
Viscosity of water $= \mu_w = 0.01$ poise

Let d_h and d_l be the sizes of the heavy and light mineral particles.
Size of the light mineral particle that settles equally with the smallest heavy mineral particle:

$$= d_l = \left(\frac{\rho_h - \rho_w}{\rho_l - \rho_w} \right)^{1/2} d_h \qquad\qquad 12.1.9$$

$$\Rightarrow d_l = \left(\frac{7.0 - 1.0}{2.0 - 1.0} \right)^{1/2} \times 10 = 105.8 \ \mu m$$

As the maximum size of the light mineral particle is 90 microns, no single particle of light mineral will settle if the upward velocity of hydraulic water is slightly less than the terminal velocity of smallest heavy mineral particle.
Terminal velocity of smallest heavy mineral particle:

$$= v_m = \frac{d_v^2 g (\rho_v - \rho_w)}{18 \mu_w} \qquad\qquad 12.1.5$$

$$\Rightarrow v_m = \frac{(0.004)^2 \, 981 (7.0 - 1.0)}{18 \times 0.01} = 0.52 \text{ cm/sec}$$

Terminal velocity of largest light mineral particle:

$$= v_m = \frac{d_g^2 g (\rho_g - \rho_w)}{18 \mu_w} \qquad\qquad 12.1.5$$

$$\Rightarrow v_m = \frac{(009)^2 \, 981 (2.0 - 1.0)}{18 \times 0.01} = 0.44 \text{ cm/sec}$$

Velocity of hydraulic water may be any value between 0.44 cm/sec and 0.52 cm/sec. Then overflow contains only light mineral particles and underflow contains only heavy mineral particles.

22.13 RECOVERY, GRADE, LOSS OF METAL ETC

Various quantities are defined in chapter 14 and formulae are indicated. The derivation of the formulae are similar to the formulae derived in the case of efficiency of the screen as detailed in chapter 22.4. The calculation of various quantities is shown in Illustration 19 and 20.

Illustration 19: *In an Iron ore concentration operation the data shown in Table 22.7 has been obtained:*

Table 22.7 Plant performance data.

	Quantity in tons	Assay value % Fe
Feed	1390	64.77
Concentrate	1112	68.08
Tailing	278	53.42

Calculate ratio of concentration, ratio of enrichment, concentrate recovery and metal recovery.
How much iron is lost in tailing?
Also calculate the metallurgical efficiency.

Solution:

Ratio of concentration $= \dfrac{F}{C} = \dfrac{1390}{1112} = 1.25$

Ratio of enrichment $= \dfrac{c}{f} = \dfrac{68.08}{64.77} = 1.05$

Concentrate recovery $= \dfrac{C}{F} \times 100 = \dfrac{1112}{1390} \times 100 = 80\%$ (also called weight recovery)

Metal recovery $= \dfrac{Cc}{Ff} \times 100 = \dfrac{1112 \times 68.08}{1390 \times 64.77} \times 100 = 84\%$

Iron lost in tailing $= \dfrac{Tt}{100} = \dfrac{278 \times 53.42}{100} = 148.5$ ton/hr

Rejection of waste in tailing $= \dfrac{T(100-t)}{F(100-f)} = \dfrac{278(100-53.42)}{1390(100-64.77)} = 0.264$

Metallurgical Efficiency $= \dfrac{R_v + J_w}{2} = \dfrac{0.84 + 0.264}{2} = 0.552 = 55.2\%$

The same formulae indicated in chapter 14 can be used to determine the tonnages of solids by considering F, C and T as weight of the total pulp in feed, concentrate and tailing and f, c and t as fraction of solids in feed, concentrate and tailing.

Illustration 20: *A cyclone is fed at the rate of 20 tons/hr of dry solids. The cyclone feed contains 30% solids, the underflow 50% solids, and the overflow 15% solids by weight. Calculate the tonnage of solids per hour in the underflow.*

Solution:
Let C and T be the weights of overflow and underflow:

Total weight of the feed $= \dfrac{20}{0.3}$ tons/hr

$C + T = F$

$\Rightarrow C + T = \dfrac{20}{0.3}$

$0.15C + 0.5T = 20$

On solving above two equations $T = 28.57$ tons/hr.
Tonnage of solids per hour in underflow $= 28.57 \times 0.5 = 14.29$ tons/hr.

References

1. **Read, H.H.:** Rutley's Elements of Mineralogy, *First Indian Edition, CBS Publishers & Distributors, Delhi, 1984,* **3.**
2. **Gaudin, A.M.:** Principles of Mineral Dressing, *TMH Edition, Tata McGraw-Hill Publishing Company Ltd., New Delhi, 2005,* **1.**
3. **Gokhale, K.V.G.K., Rao, T.C.:** Ore Deposits of India, Their Distribution and Processing, *Third Edition, Affiliated East-West Press Pvt Ltd., New Delhi, 1983,* **1.**
4. **Narayanan, C.M., Bhattacharyya, B.C.:** Mechanical Operations for Chemical Engineers, *Third Edition, Khanna Publishers, Delhi, 1999,* **1.**
5. **Bond, F.C.:** The Third Theory of Comminution, *Trans. A.I.M.E., 484, 193.*
6. **Gaudin, A.M.:** Principles of Mineral Dressing, *TMH Edition, Tata McGraw-Hill Publishing Company Ltd., New Delhi, 2005,* **25.**
7. **Gaudin, A,M., Hukki, R.T.:** Principles of comminution – Size and Surface Distribution, *A.I.M.E., 1944.*
8. **Davis, E.W.:** Fine Crushing in Ball Mills, *Trans. Am. Inst. Mining Met. Engrs., (1919), 61,* **250–296.**
9. **Stokes, G.G.:** "Mathematical and Physical Papers" (1901) also in *Trans. Cambridge Phil. Soc., 9, Part II, pp.51 et seq.* (1851).
10. **Newton, Isaac.:** "Mathematical Principles of Natural Philosophy," Book II, Trans. in to English in 1729.

Further readings

Brown, J.H.: Unit Operations in Mineral Engineering, *International Academic Services Limited, Kingston Ontario, Canada, 1979.*

Burt, R.O.: Gravity Concentration Technology, *Elsevier, Amsterdam, 1984.*

Gupta. A., Yan D.S.: Mineral Processing Design and Operation An Introduction, *Elsevier, 2006.*

Jain, S.K.: Ore Processing, *Oxford & IBH Publishing Co. Pvt. Ltd., New Delhi, 1986.*

Kelly, E.G., Spottiswood, E.J.: Introduction to Mineral Processing, *A Wiley-Interscience Publication, John Wiley & Sons, New York, 1982.*

Krishnamoorthy, K.K.: Modern Ore Testing, *Khanna Publishers, New Delhi, 1983.*

Lawrison, G.C.: Crushing and Grinding, The Size Reduction of Solid Materials, *Butterworths, London, 1974.*

Mular, A.L., Bhappu, R.B.: Mineral Processing Plant Design, *Second Edition, Society of Mining engineers, American Institute of Mining, Metallurgical, and Petroleum Engineers, Inc., New York, 1980.*

Norman L. Weiss: SME Mineral Processing Handbook, *Society of Mining Engineers of the American Institute of Mining, Metallurgical, and Petroleum Engineers, Inc., New York, 1985.*

Pryor, E.J.: Mineral Processing, *Third Edition, Applied Science Publishers Limited, London, 1974.*

Read, H.H: Principles of Mineral Dressing, *TMH Edition, Tata McGraw-Hill Publishing Company Ltd., New Delhi,* **2005, 1.**

Richards, R.H., Locke, C.E.: Textbook of Ore Dressing, *Third Edition, McGraw-Hill Book Company, Inc, New York, 1940.*

Sastri, S.R.S. et al.: Mineral Resources and Beneficiation Plant Practices in India, *Indian Institute of Mineral Engineers, Bhubaneswar.*

Subba Rao, D.V.: Coal – Its Beneficiation, *Em Kay Publications, Delhi, 2003.*

Taggart, Arthur F.: Handbook of Mineral Dressing, *A Wiley-Interscience publication, John Wiley & Sons, New York, 1945.*

Venkatachalam, S., Degaleesan, S.N.: Laboratory Experiments in Mineral Engineering, *Oxford & IBH Publishing Co., New Delhi, 1982.*

Vijayendra, H.G.: Handbook of Mineral Dressing, *Vikas Publishing House Pvt Ltd., New Delhi, 1995.*

Wills, B.A.: Mineral Processing Technology, *Fourth Edition, Pergamon Press, Oxford, 1988.*

Subject index

Printed and bound by CPI Group (UK) Ltd, Croydon, CR0 4YY

23/10/2024

01778250-0003